Design and Analysis of Gauge R&R Studies

ASA-SIAM Series on Statistics and Applied Probability

The ASA-SIAM Series on Statistics and Applied Probability is published jointly by the American Statistical Association and the Society for Industrial and Applied Mathematics. The series consists of a broad spectrum of books on topics in statistics and applied probability. The purpose of the series is to provide inexpensive, quality publications of interest to the intersecting membership of the two societies.

Burdick, R. K., Borror, C. M., and Montgomery, D. C., *Design and Analysis of Gauge R&R Studies: Making Decisions with Confidence Intervals in Random and Mixed ANOVA Models*

Albert, J., Bennett, J., and Cochran, J. J., eds., *Anthology of Statistics in Sports*

Smith, W. F., *Experimental Design for Formulation*

Baglivo, J. A., *Mathematica Laboratories for Mathematical Statistics: Emphasizing Simulation and Computer Intensive Methods*

Lee, H. K. H., *Bayesian Nonparametrics via Neural Networks*

O'Gorman, T. W., *Applied Adaptive Statistical Methods: Tests of Significance and Confidence Intervals*

Ross, T. J., Booker, J. M., and Parkinson, W. J., eds., *Fuzzy Logic and Probability Applications: Bridging the Gap*

Nelson, W. B., *Recurrent Events Data Analysis for Product Repairs, Disease Recurrences, and Other Applications*

Mason, R. L. and Young, J. C., *Multivariate Statistical Process Control with Industrial Applications*

Smith, P. L., *A Primer for Sampling Solids, Liquids, and Gases: Based on the Seven Sampling Errors of Pierre Gy*

Meyer, M. A. and Booker, J. M., *Eliciting and Analyzing Expert Judgment: A Practical Guide*

Latouche, G. and Ramaswami, V., *Introduction to Matrix Analytic Methods in Stochastic Modeling*

Peck, R., Haugh, L., and Goodman, A., *Statistical Case Studies: A Collaboration Between Academe and Industry, Student Edition*

Peck, R., Haugh, L., and Goodman, A., *Statistical Case Studies: A Collaboration Between Academe and Industry*

Barlow, R., *Engineering Reliability*

Czitrom, V. and Spagon, P. D., *Statistical Case Studies for Industrial Process Improvement*

Design and Analysis of Gauge R&R Studies

Making Decisions with Confidence Intervals in Random and Mixed ANOVA Models

Richard K. Burdick
Arizona State University
Tempe, Arizona

Connie M. Borror
University of Illinois at Urbana-Champaign
Urbana, Illinois

Douglas C. Montgomery
Arizona State University
Tempe, Arizona

siam
Society for Industrial and Applied Mathematics
Philadelphia, Pennsylvania

American Statistical Association
Alexandria, Virginia

The correct bibliographic citation for this book is as follows: Burdick, Richard K., Connie M. Borror, and Douglas C. Montgomery, *Design and Analysis of Gauge R&R Studies: Making Decisions with Confidence Intervals in Random and Mixed ANOVA Models*, ASA-SIAM Series on Statistics and Applied Probability, SIAM, Philadelphia, ASA, Alexandria, VA, 2005.

10 9 8 7 6 5 4 3 2 1

Library of Congress Cataloging-in-Publication Data

Burdick, Richard K.
Design and analysis of gauge R&R studies : making decisions with confidence intervals in random and mixed ANOVA models / Richard K. Burdick, Connie M. Borror, and Douglas c. Montgomery.
p. cm.
ISBN 0-89871-588-1 (pbk.)
1. Confidence intervals. 2. Analysis of variance. 3. Experimental design.
I. Borror, Connie M. II. Montgomery, Douglas C. III. Title.

QA276.74.B88 2005
519.5'4–dc22

2005044113

Contents

List of Figures

List of Tables

Preface

This book is written to accomplish two objectives. First, we develop a protocol for testing a measurement system. Second, we provide an up-to-date summary of methods used to construct confidence intervals in normal-based random and mixed analysis of variance (ANOVA) models.

To accomplish the first objective, we consider testing a measurement system using a gauge repeatability and reproducibility (R&R) experiment. Gauge R&R experiments use ANOVA designs to determine if a measurement system is capable of monitoring a manufacturing process. These experiments are conducted in practically every manufacturing setting and are critical in any process improvement initiative.

Although the applications in the book are specifically concerned with gauge R&R studies, the methods can be used for any application based on an ANOVA model. Thus, our second objective is to make these methods known for investigators in other fields of research. The methods we present can be used to construct confidence intervals for parameters in any random or mixed ANOVA model. We describe methods for constructing two types of confidence intervals: modified large-sample (MLS) and generalized confidence intervals (GCI). MLS intervals provide closed-form intervals that can be easily computed in a spreadsheet. The GCI method is computationally more intensive but is more general in its application than MLS. Both methods provide intervals that generally maintain the stated confidence coefficient for any sample size.

Although statistical techniques are emphasized throughout the book, an extensive background in statistics is not essential. The prerequisites are an ability to interpret a confidence interval and a basic understanding of random and mixed ANOVA models. Readers unfamiliar with these topics can gain the required background by reading Chapter 1 and Appendix A.

The book is designed as a reference, but one should read all of Chapters 1 through 4 before applying any of the methods in later chapters. Chapter 1 introduces the notation used throughout the book. Subsequent chapters consider various ANOVA models and describe how to construct confidence intervals for several model parameters. The balanced one-factor random model is described in Chapter 2. The balanced two-factor crossed random model with interaction is the topic of Chapter 3. This is an important model because it is the traditional experimental design used in gauge R&R studies. Chapter 4 presents guidelines for designing an experiment. Chapter 5 provides interval estimates for a balanced two-factor crossed random model with no interaction. Mixed two-factor designs are considered in Chapter 6. Unbalanced one- and two-factor designs are the topic of Chapter 7. Chapter 8 provides a general strategy for constructing confidence intervals in any random or mixed ANOVA design. The general results in Chapter 8 can be used to derive most of the confidence interval formulas presented in Chapters 2 through 7.

Chapter 1

Introduction

1.1 Objectives

As noted in the preface, we have two objectives for writing this book. First, we provide a protocol for testing the capability of a measurement system. Second, we provide an up-to-date summary of confidence interval estimation for variance components and means in normal-based random and mixed analysis of variance (ANOVA) models.

Although confidence intervals for ANOVA models are useful in many fields, we focus on the important application of gauge repeatability and reproducibility (R&R) studies. Confidence intervals are used in this application to determine if a measurement system is capable of monitoring a manufacturing process. This is an important application because reliable measurements are needed to monitor any manufacturing process.

Consistent with the second objective, this book extends the review of modified large-sample (MLS) confidence intervals for variance components presented in the book by Burdick and Graybill [10]. In addition, we report generalized confidence intervals (GCIs) for functions of variance components and means. GCIs are based on the process of generalized inference first proposed by Tsui and Weerahandi [65].

We hope the gauge R&R application considered in this book will be easily understood by readers in all fields of discovery. To help ensure this outcome, we provide background on this manufacturing application in this chapter. More detail on gauge R&R studies is provided in the article by Burdick, Borror, and Montgomery [8].

1.2 Gauge R&R Studies

To properly monitor and improve a manufacturing process, it is necessary to measure attributes of the process output. For any group of measurements collected for this purpose, at least part of the variation is due to the measurement system. This is because repeated measurements of any particular item do not always result in the same value. To ensure that measurement system variability is not harmfully large, it is necessary to conduct a measurement systems capability study. The purpose of this study is to

(i) determine the amount of variability in the collected data that is due to the measurement system,

(ii) isolate the sources of variability in the measurement system, and

(iii) assess whether the measurement system is suitable for use in the broader project or application.

In many measurement studies, a gauge is used to obtain replicate measurements on units by several different operators, setups, or time periods. Two components of measurement systems variability are frequently generated in such studies: repeatability and reproducibility. Repeatability represents the gauge variability when it is used to measure the same unit (with the same operator or setup or in the same time period). Reproducibility refers to the variability arising from different operators, setups, or time periods. Thus, measurement systems capability studies are often referred to as gauge repeatability and reproducibility or R&R studies. We will use this terminology throughout the book.

Gauge R&R studies are conducted in virtually every type of manufacturing industry. As noted by Spencer and Tobias [61], the Semiconductor Manufacturing Technology Consortium (SEMATECH) qualification plan requires a gauge R&R study before implementing any new process. If variation due to the measurement system is small relative to variation of the process, then the measurement system is deemed capable. This means the system can be used to monitor the process. Gauge R&R studies must be performed any time a process is modified. This is because as process variation decreases, a measurement system that was once capable may no longer be so.

Another use of a measurement system is to discriminate good product from bad product. If misclassification rates are small, then the measurement system is able to discriminate among products. While this book deals solely with a single test instance on a single product characteristic, manufacturers often use more than a single measurement on a single product characteristic to discriminate among products. Ideas for extending R&R methods to a multivariate scenario are presented by Larsen [39].

Originally, gauge R&R studies were conducted using a tabular method based on ranges and control charts. However, there are disadvantages with this tabular approach. First, confidence intervals are not easily constructed, or even possible to construct for some designs. A second disadvantage is that the range is an inefficient estimate of gauge variability. This is particularly true when sample size is moderately large. For these reasons, among others, gauge R&R studies are now analyzed using ANOVA techniques. In this book we focus on methods used to analyze ANOVA designs that include both fixed and random effects (i.e., mixed models). Readers unfamiliar with the basics of ANOVA are referred to Appendix A.

All the ANOVA models we consider are particular cases of the model

$$Y = X + E, \tag{1.1}$$

where Y is the measured value of a randomly selected part from a manufacturing process, X is the true value of the part, and E is the measurement error contributed by the measurement system. The terms X and E are independent normal random variables with means μ_P and μ_M and variances γ_P and γ_M, respectively. The subscripts P and M denote process and measurement system, respectively. These assumptions imply that Y has a normal distribution with mean $\mu_Y = \mu_P + \mu_M$ and variance $\gamma_Y = \gamma_P + \gamma_M$. Additionally, the

Table 1.1. *Parameters in a gauge R&R study.*

Symbol	Definition
μ_Y	Mean of population of measurements
γ_P	Variance of the process
γ_M	Variance of the measurement system
$\gamma_R = \dfrac{\gamma_P}{\gamma_M}$	Ratio of process variance to measurement variance

covariance between Y and X is γ_P and the correlation between Y and X is $\sqrt{\gamma_P/(\gamma_P + \gamma_M)}$. Note that as the measurement system improves and γ_M decreases, the correlation between the measured value and the true value approaches one.

The mean μ_M is referred to as the bias of the measurement system. Typically, this bias can be eliminated by proper calibration of the system. Thus, we assume $\mu_M = 0$ throughout the book so that $\mu_Y = \mu_P$. If this assumption is violated, it will affect estimation of μ_P but not estimation of the variances. Since the variances are of primary interest in a gauge R&R study, this assumption is merely for convenience of notation.

Table 1.1 reports the parameters in model (1.1) that are the focus of this book. It is useful to describe these parameters in the context of an example. Houf and Berman [32] describe a process used to manufacture a power module for a line of motor starters. An important attribute is the thermal performance of the power module. Measurements of thermal performance are taken on a sample of power modules using a thermal resistance measuring instrument. The units of measurement are C° per watt. Variability in the data can be attributed to

(i) the manufacturing process used to produce the power modules (γ_P) and

(ii) measurement error contributed by the thermal resistance measuring instrument (γ_M).

The purpose of the gauge R&R study is to determine if the variability due to source (ii) is small relative to source (i). Thus, it is common to examine the ratio of these two sources of variability, γ_R. A measurement system is capable if γ_R is sufficiently large. Operational definitions for large are described in the next section.

1.3 Capability Measures

Several criteria are used to determine if a measurement system is capable. All of these criteria are based on the parameters shown in Table 1.1.

The precision-to-tolerance ratio (PTR) is

$$\text{PTR} = \frac{k\sqrt{\gamma_M}}{USL - LSL}, \tag{1.2}$$

where USL and LSL are specification limits and k is either 5.15 or 6. The value $k = 6$ corresponds to the number of standard deviations between the "natural" tolerance limits that contain the middle 99.73% of a normal process. The value $k = 5.15$ corresponds to limits that contain the middle 99% of a normal population.

The Automotive Industry Action Group (AIAG) Measurement Systems Analysis manual [3, p. 77] recommends the following rule of thumb for using PTR to determine the capability of a measurement system:

(i) PTR ≤ 0.1: the measurement system is capable.

(ii) $0.1 <$ PTR ≤ 0.3: the measurement system may be capable depending on factors such as process capability and costs of misclassification.

(iii) PTR > 0.3: the measurement system is not capable.

Montgomery and Runger [53] and Mader, Prins, and Lampe [48] note that PTR does not necessarily give a good indication of how well a measurement system performs for a particular process. This is because a process with a high capability can tolerate a measurement system with a higher PTR than a process that is not as capable. For this reason, the capability of a measurement system is often determined by functions of the ratio γ_R.

The signal-to-noise ratio (SNR) is a function of γ_R defined as

$$\text{SNR} = \sqrt{2\gamma_R}. \tag{1.3}$$

The AIAG Measurement Systems Analysis manual [3, p. 117] defines SNR as the number of distinct categories that can be reliably distinguished by the measurement system. A value of five or greater is recommended, and a value of less than two indicates the measurement system is of no value in monitoring the process.

Another function of γ_R is the intraclass correlation

$$\rho = \frac{\gamma_R}{1+\gamma_R} = \frac{\gamma_P}{\gamma_P + \gamma_M}. \tag{1.4}$$

This parameter represents the correlation between two measurements taken on the same part. It is also interpreted as the proportion of total variation in the measurements due to the process.

The criteria described above are the most popular criteria for evaluating the capability of a measurement system. However, given that a measurement system is designed to discriminate between good and bad products, we believe a more meaningful criterion is the probability of part misclassification. A method for evaluating a measurement system using this criterion is described in Section 1.6.

Although the primary reason for conducting a gauge R&R study is to study the measurement system, one may also want to estimate the capability of the monitored process. One common measure of process capability is

$$C_p = \frac{USL - LSL}{6\sqrt{\gamma_P}}. \tag{1.5}$$

In addition to providing interval estimates for PTR and SNR, we also provide confidence intervals for C_p.

1.4 Confidence Intervals

It is never possible to know the exact values of the parameters in Table 1.1. Rather, they are estimated using the data collected in the gauge R&R experiment. Confidence intervals are used to quantify the uncertainty associated with the estimation process. For demonstration, consider the parameter γ_M. A $100(1-\alpha)\%$ confidence interval for γ_M is a random interval with a lower bound L and an upper bound U that are functions of the sample values such that $\Pr[L \leq \gamma_M \leq U] = 1-\alpha$. The term $100(1-\alpha)\%$ is called the confidence coefficient and is selected before data collection. Typical values for the confidence coefficient are 90%, 95%, and 99%.

A confidence interval that satisfies $\Pr[L \leq \gamma_M \leq U] = 1-\alpha$ is called an exact two-sided confidence interval. Often such exact intervals do not exist, and $\Pr[L \leq \gamma_M \leq U]$ is only approximately equal to $1-\alpha$. An approximate interval that has a realized confidence coefficient greater than the stated level (i.e., $\Pr[L \leq \gamma_M \leq U] > 1-\alpha$) is called a conservative interval. An approximate interval that has a realized confidence coefficient of less than the stated level (i.e., $\Pr[L \leq \gamma_M \leq U] < 1-\alpha$) is called a liberal interval. In general, conservative intervals are preferred when only approximate intervals are available. Liberal intervals can be recommended if the confidence coefficient is not too much less than the stated level.

We provide confidence intervals for the parameters in Table 1.1 for each design considered in this book. Transformations can be applied to form intervals for other functions of these parameters. To demonstrate, suppose $[L, U]$ is an exact 95% two-sided confidence interval for γ_R. This means

$$\Pr[L \leq \gamma_R \leq U] = 0.95.$$

Thus

$$\begin{aligned}
\Pr[L \leq \gamma_R \leq U] &= \Pr\left[\frac{1}{L} \geq \frac{1}{\gamma_R} \geq \frac{1}{U}\right] \\
&= \Pr\left[\frac{1}{L} + 1 \geq \frac{1}{\gamma_R} + 1 \geq \frac{1}{U} + 1\right] \\
&= \Pr\left[\frac{1+L}{L} \geq \frac{1+\gamma_R}{\gamma_R} \geq \frac{1+U}{U}\right] \\
&= \Pr\left[\frac{L}{1+L} \leq \frac{\gamma_R}{1+\gamma_R} \leq \frac{U}{1+U}\right] \\
&= 0.95.
\end{aligned}$$

So from the definition of ρ shown in Equation (1.4), the 95% confidence interval for ρ is

$$\left[\frac{L}{1+L}, \frac{U}{1+U}\right].$$

There are few instances in which exact confidence intervals are available for the parameters in Table 1.1. When exact confidence intervals do not exist, two approaches are commonly used for constructing approximate confidence intervals. The first approach is based on the MLS methods first proposed by Graybill and Wang [25] and summarized in the book by Burdick and Graybill [10]. This approach provides closed-form intervals. The second approach is based on a computer-intensive method referred to as generalized confidence intervals, or GCI. Tsui and Weerahandi [65] introduced the concept of generalized inference for testing hypotheses when exact methods do not exist. Weerahandi [71] extended this concept to construct GCIs. This method was applied to a gauge R&R two-factor study by Hamada and Weerahandi [27]. Chiang [14] proposed a method called surrogate variables that produces the same intervals (see Iyer and Mathew [33]). Construction of a GCI requires a generalized pivotal quantity (GPQ) with a distribution that is free of the parameters under study. Approximate confidence intervals are then constructed by computing percentiles of the GPQ using either numerical integration or simulation. We demonstrate this process in Chapter 2 and report GPQs for each model examined in later chapters. More details concerning the MLS and GCI approaches are presented in Appendix B. Chapter 8 presents a general strategy for using these methods to construct confidence intervals in any random or mixed ANOVA model.

We note that the mixed models considered in this book can also be analyzed using the MIXED procedure of SAS. However, the resulting likelihood-based intervals are valid only for large samples. We do not recommend using MIXED to compute confidence intervals for functions of variance components. Burdick and Larsen [11] demonstrate that intervals based on the asymptotic standard errors of REML estimates fail to maintain confidence levels in typical gauge R&R studies. Additionally, Park and Burdick [56] demonstrate that the CL = Wald option used in MIXED produces upper bounds that are much greater than the bounds recommended in this book.

Empirical comparisons of MLS and GCI suggest the approaches provide comparable results in most cases. Since the MLS intervals are closed form, they are particularly amenable for computations in a spreadsheet program. An advantage of GCI is that it offers a general approach that can be used for any ANOVA model.

1.5 R&R Graphs

A useful graph for evaluating the capability of a measurement system is the R&R graph. This graph displays confidence intervals for PTR and SNR as defined in Equations (1.2) and (1.3). Consider an application where it is desired to have PTR $\leq \phi_1$ and SNR $\geq \phi_2$, where $\phi_1 = 0.3$ and $\phi_2 = 5$. These inequalities are represented graphically in Figure 1.1. The two lines divide the graph into four regions.

(i) Region 1: the measurement system satisfies both criteria.

(ii) Region 2: the measurement system satisfies the SNR criterion but not the PTR criterion.

(iii) Region 3: the measurement system does not meet either criterion.

(iv) Region 4: the measurement system satisfies the PTR criterion but not the SNR criterion.

Once the confidence intervals for PTR and SNR have been computed, they are displayed on the grid in Figure 1.1. By noting the regions covered by the intersection of the confidence

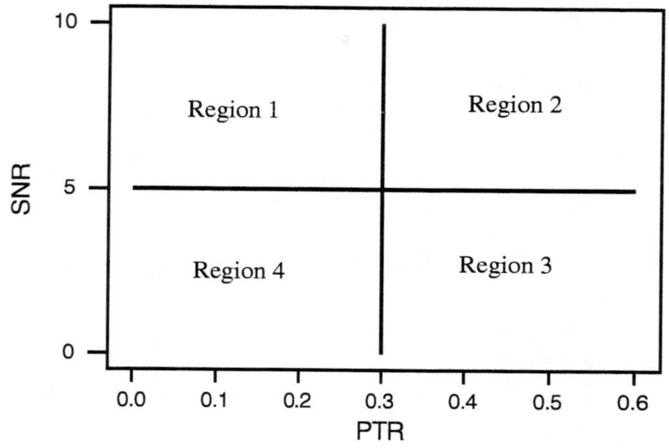

Figure 1.1. *R&R graph.*

intervals, one can draw conclusions concerning the capability of the measurement system. Majeske and Andrews [49] recommended similar graphs in the space of γ_P and γ_M.

1.6 Misclassification Rates

As noted in Section 1.3, the quality of a measurement system can be defined by how well it discriminates between good and bad parts. To explain this concept, we begin with the model in Equation (1.1). Recall that Y is the measured value of a part and X is the true value. The joint probability density function of Y and X is bivariate normal with mean vector $[\mu_Y \quad \mu_P]'$ and covariance matrix

$$\begin{bmatrix} \gamma_P + \gamma_M & \gamma_P \\ \gamma_P & \gamma_P \end{bmatrix}. \tag{1.6}$$

We represent this density function as $f(y, x)$.

A manufactured part is in conformance if

$$LSL \leq X \leq USL, \tag{1.7}$$

where LSL and USL are the lower and upper specification limits, respectively. A measurement system will pass a part if

$$LSL \leq Y \leq USL. \tag{1.8}$$

If (1.7) is true and (1.8) is false, then a conforming part is misclassified as a failure. This is called a false failure (FF). Alternatively, if (1.7) is false but (1.8) is true, a failure is misclassified as a good part. This is called a missed fault (MF). Figure 1.2 illustrates the regions of FF and MF on an equal density contour of the bivariate normal distribution. Probabilities of the misclassification errors FF and MF are computed as either joint or

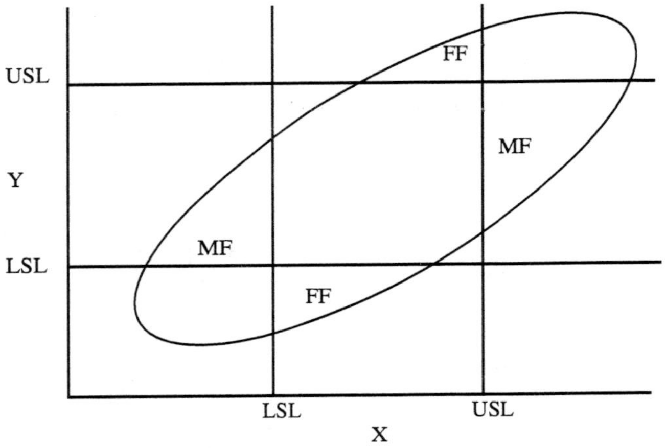

Figure 1.2. *MF and FF regions of an equal density contour.*

conditional probabilities. To demonstrate, a false failure occurs when (1.7) is true and (1.8) is false. Representing this joint probability with the symbol δ,

$$\begin{aligned}\delta &= \Pr[(LSL \le X \le USL) \text{ and } (Y < LSL \text{ or } Y > USL)] \\ &= \Pr[(LSL \le X \le USL \text{ and } Y < LSL) \text{ or} \\ &\quad (LSL \le X \le USL \text{ and } Y > USL)] \\ &= \int_{LSL}^{USL} \int_{-\infty}^{LSL} f(y,x)dydx + \int_{LSL}^{USL} \int_{USL}^{\infty} f(y,x)dydx. \end{aligned} \tag{1.9}$$

A missed fault occurs when (1.7) is false and (1.8) is true. If we represent this joint probability with the symbol β,

$$\begin{aligned}\beta &= \Pr[(X < LSL \text{ or } X > USL) \text{ and } (LSL \le Y \le USL)] \\ &= \Pr[(X < LSL \text{ and } LSL \le Y \le USL) \text{ or} \\ &\quad (X > USL \text{ and } LSL \le Y \le USL)] \\ &= \int_{-\infty}^{LSL} \int_{LSL}^{USL} f(y,x)dydx + \int_{USL}^{\infty} \int_{LSL}^{USL} f(y,x)dydx. \end{aligned} \tag{1.10}$$

The probability δ is referred to as producer's risk and the probability β is referred to as consumer's risk. Some investigators prefer to compute the conditional probabilities

$$\delta_c = \frac{\delta}{\pi} \tag{1.11}$$

and

$$\beta_c = \frac{\beta}{1-\pi}, \tag{1.12}$$

where

$$\pi = \int_{LSL}^{USL} f(x)dx \tag{1.13}$$

is the probability of a good part and $f(x)$ is the marginal probability density function for X which is normal with mean μ_p and variance γ_p. That is, δ_c is the conditional probability that a part fails given the part is good. The symbol β_c is the conditional probability that a part passes given the part is bad.

When considering whether misclassification rates are sufficiently small, it is useful to compare them to what one could obtain by chance. To demonstrate, suppose the value of π in Equation (1.13) is known. A chance process for classifying parts is to take no measurements and randomly classify π of the parts as good and $1 - \pi$ as bad. For this chance process, the probability of a false failure is

$$\begin{aligned}\delta^* &= \Pr[(LSL \le X \le USL) \text{ and (Part classified as bad)}] \\ &= \Pr(LSL \le X \le USL) \times \Pr(\text{Part classified as bad}) \\ &= \pi(1-\pi).\end{aligned}$$

Likewise, the probability of a missed fault for this chance process is

$$\begin{aligned}\beta^* &= \Pr[(X < LSL \text{ or } X > USL) \text{ and (Part classified as good)}] \\ &= \Pr(X < LSL \text{ or } X > USL) \times \Pr(\text{Part classified as good}) \\ &= (1-\pi)\pi.\end{aligned}$$

Thus, for a measurement system to be useful, the misclassification rates δ and β should be less than what one could get by chance, δ^* and β^*. One can make a similar argument using the conditional probabilities δ_c and β_c. In this case, the conditional probabilities that could be obtained by chance are $\delta_c^* = 1 - \pi$ and $\beta_c^* = \pi$.

To facilitate comparisons of the misclassification rates to the chance criteria, we will compute ratios of the misclassification rates to the chance rates. If a measurement system is able to discriminate better than the chance process, these indexes will be less than one. These ratios are defined as

$$\begin{aligned}\delta_{index} &= \frac{\delta}{\delta^*} \\ &= \frac{\delta}{\pi(1-\pi)}\end{aligned} \tag{1.14}$$

and

$$\begin{aligned}\beta_{index} &= \frac{\beta}{\beta^*} \\ &= \frac{\beta}{(1-\pi)\pi}.\end{aligned} \tag{1.15}$$

Notice $\delta_c/\delta_c^* = \delta_{index}$ and $\beta_c/\beta_c^* = \beta_{index}$, so these indexes can be used for both the unconditional and the conditional definitions of misclassification rates. We will use these ratios and the capability measures described in Section 1.3 to determine the capability of a measurement system.

The misclassification indexes in Equations (1.14) and (1.15) can be computed for given values of μ_Y, μ_P, γ_P, γ_M, LSL, and USL by integrating bivariate normal distributions as shown in Equations (1.9) and (1.10). More typically, these parameters are unknown and must be estimated from the measurements obtained in the R&R study. Since we are assuming $\mu_Y = \mu_P$, the overall sample mean provides an estimate of both μ_Y and μ_P. If there is a systematic bias in the system, then it is necessary to estimate μ_P from another source of information. Such an estimate might be based on historical data from the process. Estimates for γ_P and γ_M will be available from the measurement data even if systematic bias is present in the measurement system.

To account for the uncertainty in the estimates of μ_Y, μ_P, γ_P, and γ_M, we will compute generalized confidence intervals for δ_{index} and β_{index} as well as for the misclassification rates in Equations (1.9)–(1.12). The method used to perform these calculations is detailed by Burdick et al. [13]. Simply, the calculations are performed by replacing the parameters μ_Y, μ_P, γ_P, and γ_M with GPQs. These pivotal quantities will be reported in each chapter for the specific design under consideration.

1.7 Determination of *F*-Values

All the MLS confidence interval formulas require percentiles from either the chi-squared or F-distribution. The symbol $\chi^2_{\alpha:df}$ represents the percentile of a chi-squared distribution with df degrees of freedom and area α to the *left*. The symbol $F_{\alpha:df1,df2}$ represents the percentile of an F-distribution with $df1$ and $df2$ degrees of freedom and area α to the *left*. Since percentiles from these distributions are related by the equations

$$\frac{\chi^2_{\alpha:df}}{df} = F_{\alpha:df,\infty} = \frac{1}{F_{1-\alpha:\infty,df}}, \tag{1.16}$$

all required values can be obtained from an F-table. Appendix C provides F-tables that can be used while working the example problems in this book. To practice, let $\alpha = 0.025$ and verify from Table C.3 that $F_{.025,3,10} = 0.0694$.

In practice, it might be more convenient to use a software program to determine F-values. The F-value with $\alpha = 0.025$, $df1 = 3$, and $df2 = 10$ can be found using Excel 2000 by typing the formula "$= FINV(.975, 3, 10)$" into a blank cell of a worksheet. The general form of this function is $FINV(1 - \alpha, df1, df2)$. The general function in SAS that returns this value is $FINV(\alpha, df1, df2)$. Making use of Equation (1.16), the value $F_{.025:3,\infty} = 0.0719$ can be computed in Excel by typing "$= CHIINV(.975, 3)/3$." The appropriate statement in SAS is "$CINV(.025, 3)/3$."

1.8 Computer Programs

Example data sets, SAS code, and Excel spreadsheets used to perform the example computations in this book are available at www.siam.org/books/sa17. These programs can be used with only a modest knowledge of either Excel or SAS.

Excel spreadsheets are provided for all the MLS intervals presented in the book. The GCIs are not computed in Excel. The user must input mean squares and degrees of freedom into the Excel spreadsheet. Since Excel has a limited number of ANOVA programs, it likely will be necessary to compute these values using another software package.

The SAS programs are self-contained and require input of the raw data. The programs compute the appropriate ANOVA table and use this information to construct both the MLS and the generalized confidence intervals. There is a separate SAS program for each model considered in the book. Detailed instructions for using both sets of programs are provided at www.siam.org/books/sa17.

1.9 Summary

This chapter stated the book objectives and provided a brief description of gauge R&R studies. Each chapter of the book considers a gauge R&R study defined by a different ANOVA model. MLS and generalized confidence intervals are presented for each parameter in Table 1.1 and for the capability measures described in Section 1.3. Confidence intervals for the misclassification rates described in Section 1.6 are computed using generalized inference. Confidence intervals for other parameters are presented in some of the chapters.

It is recommended that the reader complete Chapters 1 through 4 before using other material in the book. Chapter 2 presents important notation, Chapter 3 considers the most typical gauge R&R design, and Chapter 4 provides information concerning experimental design. Chapter 8 contains general results that are used to derive most of the confidence intervals in Chapters 2 through 7. A reader interested in the derivations may wish to start by reading Chapters 1 and 8.

Each chapter of the book includes a table that defines the notation used in that chapter. The reader is cautioned that definitions sometimes change from chapter to chapter, and it is necessary to refer to the appropriate table of definitions before using any of the reported confidence intervals.

Chapter 2

Balanced One-Factor Random Models

2.1 Introduction

The first model we consider is the balanced one-factor random design. This model describes a gauge R&R study in which a single operator selects a random sample of p parts and measures each part r times using the same measurement gauge.

Table 2.1 presents a subset of data reported by Houf and Berman [32]. In this example, the monitored parts are power modules for a line of motor starters. The response is the thermal performance of the module measured in C° per watt. Measurements are taken on the device using a thermal resistance measuring instrument. The data shown in the table represent measurements taken by a single operator. Each response has been multiplied by 100 for convenience of scale. The $p = 10$ parts were sampled at random from the manufacturing process, and $r = 2$ replicate measures were made on each part.

We now present the model used to analyze the data in Table 2.1. The completed analysis is reported in Section 2.5.

Table 2.1. *Example data for balanced one-factor random model.*

Part	Measurements ($r = 2$)
1	37, 38
2	42, 41
3	30, 31
4	42, 43
5	28, 30
6	42, 42
7	25, 26
8	40, 40
9	25, 25
10	35, 34

Table 2.2. *ANOVA for model* (2.1).

Source of variation	Degrees of freedom	Mean square	Expected mean square
Parts	$p-1$	S_P^2	$\theta_P = \sigma_E^2 + r\sigma_P^2$
Replicates	$p(r-1)$	S_E^2	$\theta_E = \sigma_E^2$

Table 2.3. *Mean squares and means for model* (2.1).

Statistic	Definition
S_P^2	$\dfrac{r\Sigma_i(\overline{Y}_{i*} - \overline{Y}_{**})^2}{p-1}$
S_E^2	$\dfrac{\Sigma_i\Sigma_j(Y_{ij} - \overline{Y}_{i*})^2}{p(r-1)}$
$\overline{Y}_{i*}$	$\dfrac{\Sigma_j Y_{ij}}{r}$
$\overline{Y}_{**}$	$\dfrac{\Sigma_i\Sigma_j Y_{ij}}{pr}$

2.2 The Model

The balanced one-factor random model is

$$Y_{ij} = \mu_Y + P_i + E_{ij}, \qquad (2.1)$$
$$i = 1, \ldots, p, \quad j = 1, \ldots, r,$$

where μ_Y is a constant, and P_i and E_{ij} are jointly independent normal random variables with means of zero and variances σ_P^2 and σ_E^2, respectively. The ANOVA for model (2.1) is shown in Table 2.2, and definitions for mean squares and means are shown in Table 2.3.

Table 2.4 reports distributional results based on the assumptions of model (2.1). These results are needed to derive the confidence intervals presented in this chapter.

Table 2.5 defines the parameters from Table 1.1 in terms of the notation for model (2.1). Point estimators for the parameters are included in this table. The estimators $\widehat{\mu}_Y$, $\widehat{\gamma}_P$, and $\widehat{\gamma}_M$ are minimum variance unbiased (MVU) estimators.

2.3 MLS Confidence Intervals

We now present MLS confidence intervals for the parameters in Table 2.5. All these intervals are special cases of the general results presented in Chapter 8. Each interval requires constants that are functions of F-values. For convenience, Table 2.6 summarizes these constants for the one-factor design where $F_{\alpha:df1,df2}$ represents the F-value with degrees of

Table 2.4. *Distributional results for model* (2.1).

Result	
1	$\overline{Y}_{**}$, S_P^2, and S_E^2 are jointly independent.
2	$(p-1)S_P^2/\theta_P$ is a chi-squared random variable with $p-1$ degrees of freedom.
3	$p(r-1)S_E^2/\theta_E$ is a chi-squared random variable with $p(r-1)$ degrees of freedom.
4	$\overline{Y}_{**}$ is a normal random variable with mean μ_Y and variance $\dfrac{\theta_P}{pr}$.

Table 2.5. *Gauge R&R parameters and point estimators for model* (2.1).

Gauge R&R notation	Model (2.1) representation	Point estimator
μ_Y	μ_Y	$\widehat{\mu}_Y = \overline{Y}_{**}$
γ_P	σ_P^2	$\widehat{\gamma}_P = \dfrac{S_P^2 - S_E^2}{r}$
γ_M	σ_E^2	$\widehat{\gamma}_M = S_E^2$
γ_R	$\dfrac{\sigma_P^2}{\sigma_E^2}$	$\widehat{\gamma}_R = \dfrac{S_P^2/S_E^2 - 1}{r}$

Table 2.6. *Constants used in confidence intervals for model* (2.1). *Values are for* $\alpha = 0.05$, $p = 10$, *and* $r = 2$.

Constant	Definition	Value
G_1	$1 - F_{\alpha/2:\infty,p-1}$	0.5269
G_2	$1 - F_{\alpha/2:\infty,p(r-1)}$	0.5118
H_1	$F_{1-\alpha/2:\infty,p-1} - 1$	2.333
H_2	$F_{1-\alpha/2:\infty,p(r-1)} - 1$	2.080
F_1	$F_{1-\alpha/2:p-1,p(r-1)}$	3.7790
F_2	$F_{\alpha/2:p-1,p(r-1)}$	0.2523
G_{12}	$\dfrac{(F_1-1)^2 - G_1^2F_1^2 - H_2^2}{F_1}$	−0.1501
H_{12}	$\dfrac{(1-F_2)^2 - H_1^2F_2^2 - G_2^2}{F_2}$	−0.1951

freedom $df1$ and $df2$ and area α to the *left*. For illustration, we report values that provide a two-sided confidence interval with a 95% confidence coefficient ($\alpha = 0.05$) for a design with $p = 10$ parts and $r = 2$ replicates. All these values can be verified using Table C.3 in Appendix C. For example, with $\alpha = 0.05$ and $p = 10$, $F_{\alpha/2:\infty,p-1} = F_{0.025:\infty,9} = 0.4731$ and so $G_1 = 1 - 0.4731 = 0.5269$.

2.3.1 Interval for μ_Y

The confidence interval for μ_Y is based on Results 1, 2, and 4 of Table 2.4. Using these results, $(\overline{Y}_{**} - \mu_Y)/\sqrt{S_P^2/(pr)}$ has a t-distribution with $p-1$ degrees of freedom. From this fact, the bounds of an exact $100(1-\alpha)\%$ confidence interval for μ_Y are

$$L = \overline{Y}_{**} - \sqrt{\frac{S_P^2 F_{1-\alpha:1,p-1}}{pr}}$$

and

$$U = \overline{Y}_{**} + \sqrt{\frac{S_P^2 F_{1-\alpha:1,p-1}}{pr}}. \tag{2.2}$$

Here $t_{1-\alpha/2:p-1}$ is represented as $\sqrt{F_{1-\alpha:1,p-1}}$ (see, e.g., [23, p. 66]). For example, if $\alpha = 0.05$ and $p = 10$, then $t_{0.975:9} = \sqrt{F_{0.95:1,9}} = \sqrt{5.1174} = 2.262$.

2.3.2 Interval for γ_P

The parameter γ_P can be written as a difference of the two expected mean squares in Table 2.2. Since $\gamma_P = \sigma_P^2$, we can write $\gamma_P = (\theta_P - \theta_E)/r$. Burdick and Graybill [10, pp. 37–38] review several methods for constructing confidence intervals on a difference of expected mean squares. One method they recommend is the MLS procedure proposed by Ting et al. [64]. Using this method, the bounds of an approximate $100(1-\alpha)\%$ confidence interval for γ_P are

$$L = \widehat{\gamma}_P - \frac{\sqrt{V_{LP}}}{r}$$

and

$$U = \widehat{\gamma}_P + \frac{\sqrt{V_{UP}}}{r}, \tag{2.3}$$

where

$$V_{LP} = G_1^2 S_P^4 + H_2^2 S_E^4 + G_{12} S_P^2 S_E^2,$$
$$V_{UP} = H_1^2 S_P^4 + G_2^2 S_E^4 + H_{12} S_P^2 S_E^2,$$

$\widehat{\gamma}_P$ is defined in Table 2.5, and G_1, G_2, H_1, H_2, G_{12}, and H_{12} are defined in Table 2.6.

Another recommended interval was proposed independently by Tukey [66] and Williams [73]. This approximate $100(1-\alpha)\%$ confidence interval for γ_P is

$$L = \frac{(S_P^2 - S_E^2 F_1)(1 - G_1)}{r}$$

and

$$U = \frac{(S_P^2 - S_E^2 F_2)(1 + H_1)}{r}, \tag{2.4}$$

where G_1, H_1, F_1, and F_2 are defined in Table 2.6.

Intervals computed from Equations (2.3) and (2.4) will be very similar. Both generally maintain the stated confidence level, and the interval lengths are comparable. The lower bound of each interval is negative if $S_P^2/S_E^2 < F_1$. Since $\gamma_P \geq 0$, negative bounds are increased to zero. This will not affect the confidence coefficient since all eliminated values from the interval are negative.

2.3.3 Interval for γ_M

Since $\gamma_M = \sigma_E^2$, Result 3 in Table 2.4 is used to obtain the exact $100(1-\alpha)\%$ confidence interval for σ_E^2,

$$L = (1 - G_2)S_E^2$$

and

$$U = (1 + H_2)S_E^2, \tag{2.5}$$

where G_2 and H_2 are defined in Table 2.6. Here the chi-squared value $\chi^2_{\alpha/2:p(r-1)}$ is represented as $p(r-1)F_{\alpha/2:p(r-1),\infty}$. (See Equation (1.16).)

2.3.4 Interval for γ_R

The parameter γ_R is represented as

$$\begin{aligned}\gamma_R &= \frac{\gamma_P}{\gamma_M} \\ &= \frac{\theta_P - \theta_E}{r\theta_E} \\ &= \frac{\theta_P}{r\theta_E} - \frac{1}{r}.\end{aligned}$$

Results 1–3 in Table 2.4 imply $(S_P^2\theta_E)/(S_E^2\theta_P)$ has an F-distribution with $p-1$ and $p(r-1)$ degrees of freedom. Thus, the bounds of an exact $100(1-\alpha)\%$ confidence interval for γ_R are

$$L = \frac{S_P^2}{rS_E^2 F_1} - \frac{1}{r}$$

and

$$U = \frac{S_P^2}{rS_E^2 F_2} - \frac{1}{r}, \tag{2.6}$$

where F_1 and F_2 are defined in Table 2.6. Since $\gamma_R \geq 0$, negative values of L and U are set to zero.

2.4 Generalized Confidence Intervals

Generalized confidence intervals, or GCIs, were briefly discussed in Section 1.4. The GCIs for μ_Y, γ_M, and γ_R all reduce to the exact intervals provided in the previous section. Thus, we need only consider a GCI for γ_P. The GPQ for γ_P is

$$\text{GPQ}(\gamma_P) = \max\left[0, \frac{(p-1)s_P^2}{rW_1} - \frac{p(r-1)s_E^2}{rW_2}\right], \tag{2.7}$$

where s_P^2 and s_E^2 are the realized values of S_P^2 and S_E^2 for a given data set, and W_1 and W_2 are independent chi-squared random variables with $p-1$ and $p(r-1)$ degrees of freedom, respectively. (Appendix B describes the approach used to derive this GPQ.)

We use the following process to construct a GCI for γ_P:

1. Compute S_P^2 and S_E^2 for the collected data and denote the realized values as s_P^2 and s_E^2, respectively.

2. Simulate N values of the GPQ shown in Equation (2.7) by simulating N independent values each of W_1 and W_2. If possible, we recommend that N be at least 100,000.

3. Order the N GPQ values obtained in step 2 from least to greatest.

4. Define the lower bound for a $100(1-\alpha)\%$ interval as the value in position $N \times (\alpha/2)$ of the ordered set in step 3. Define the upper bound as the value in position $N \times (1-\alpha/2)$ of this same ordered set.

As discussed in Section 1.6, GCIs can be computed for the producer's risk (δ), the consumer's risk (β), δ_c, β_c, δ_{index}, and β_{index}. This procedure is detailed by Burdick et al. [13]. In brief, one computes the misclassification rates defined in Equations (1.9)–(1.15) by replacing the unknown parameters with their respective GPQs. These GPQs are shown in Table 2.7, where $\overline{y}_{**}$, s_P^2, and s_E^2 are the realized values of $\overline{Y}_{**}$, S_P^2, and S_E^2, respectively, Z is a normal random variable with mean zero and variance one, and ϵ is a small positive value. Notice the GPQ for γ_P has a minimum value of ϵ, whereas the minimum value in Equation (2.7) is zero. This change is made so that the integrals used to compute the misclassification rates are always defined. We set $\epsilon = 0.001$ in all the examples presented in this book.

To demonstrate, the algorithm used to compute a generalized confidence interval for δ is as follows:

1. Compute $\overline{Y}_{**}$, S_P^2, and S_E^2 for the collected data and denote the realized values as $\overline{y}_{**}$, s_P^2, and s_E^2, respectively.

2. Simulate N values of each GPQ shown in Table 2.7 by simulating N independent values each of W_1, W_2, and Z. We use $N = 100{,}000$ in the examples that follow.

3. Compute N values of δ using Equation (1.9) by replacing μ_Y, μ_P, $\gamma_P + \gamma_M$, and γ_P with the GPQs formed in step 2.

4. Order the N values obtained in step 3 from least to greatest.

Table 2.7. *GPQs for misclassification rates in model* (2.1).

Parameter	GPQ
$\mu_Y = \mu_P$	$\bar{y}_{**} - Z\sqrt{\frac{(p-1)s_P^2}{prW_1}}$
$\gamma_P + \gamma_M$	$\frac{(p-1)s_P^2}{rW_1} + \frac{p(r-1)^2 s_E^2}{rW_2}$
γ_P	$\max\left[\epsilon, \frac{(p-1)s_P^2}{rW_1} - \frac{p(r-1)s_E^2}{rW_2}\right]$

Table 2.8. *ANOVA for example one-factor model.*

Source of variation	Degrees of freedom	Mean square
Parts	9	94.47
Replicates	10	0.50

5. Define the lower bound for a $100(1-\alpha)\%$ interval as the value in position $N \times (\alpha/2)$ of the ordered set in step 4. Define the upper bound as the value in position $N \times (1-\alpha/2)$ of this same ordered set.

This same process is used to compute confidence intervals for all the other misclassification rates. We now present a numerical example to demonstrate these computations.

2.5 Numerical Example

In this section, we use the data in Table 2.1 to demonstrate the computations presented in the previous two sections. We also construct confidence intervals for PTR, SNR, and C_p.

The computed ANOVA is shown in Table 2.8. The overall sample average is 34.8. To construct two-sided 95% confidence intervals, we use the constants shown in Table 2.6, where $\alpha = 0.05$, $p = 10$, and $r = 2$.

2.5.1 Interval for μ_Y

The lower and upper confidence bounds for μ_Y are computed using Equation (2.2). With $\alpha = 0.05$, $F_{1-\alpha:1,p-1} = F_{0.95:1,9} = 5.1174$ and the 95% confidence interval is

$$L = \overline{Y}_{**} - \sqrt{\frac{S_P^2 F_{1-\alpha:1,p-1}}{pr}}$$
$$= 34.8 - \sqrt{\frac{94.47(5.1174)}{10(2)}}$$
$$= 29.9$$

and

$$U = \overline{Y}_{**} + \sqrt{\frac{S_P^2 F_{1-\alpha:1,p-1}}{pr}}$$
$$= 34.8 + \sqrt{\frac{94.47(5.1174)}{10(2)}}$$
$$= 39.7.$$

2.5.2 Intervals for γ_P

The point estimate for γ_P is

$$\widehat{\gamma}_P = \frac{S_P^2 - S_E^2}{r}$$
$$= \frac{94.47 - 0.50}{2}$$
$$= 47.0.$$

The MLS confidence interval is shown in Equation (2.3). In this example,

$$V_{LP} = G_1^2 S_P^4 + H_2^2 S_E^4 + G_{12} S_P^2 S_E^2$$
$$= (.5269)^2(94.47)^2 + (2.080)^2(0.50)^2 + (-0.1501)(94.47)(0.50)$$
$$= 2.47 \times 10^3$$

and

$$V_{UP} = H_1^2 S_P^4 + G_2^2 S_E^4 + H_{12} S_P^2 S_E^2$$
$$= (2.333)^2(94.47)^2 + (0.5118)^2(0.50)^2 + (-0.1951)(94.47)(0.50)$$
$$= 4.86 \times 10^4.$$

Substituting this information into Equation (2.3) yields the 95% confidence interval

$$L = 47.0 - \frac{\sqrt{2.47 \times 10^3}}{2} = 22.1$$

and

$$U = 47.0 + \frac{\sqrt{4.86 \times 10^4}}{2} = 157. \tag{2.8}$$

The Tukey–Williams interval shown in Equation (2.4) is

$$L = \frac{(S_P^2 - S_E^2 F_1)(1 - G_1)}{r}$$
$$= \frac{(94.47 - 0.50(3.7790))(1 - 0.5269)}{2}$$
$$= 21.9$$

and

$$U = \frac{(S_P^2 - S_E^2 F_2)(1 + H_1)}{r}$$
$$= \frac{(94.47 - 0.50(0.2523))(1 + 2.333)}{2}$$
$$= 157.$$

This interval is comparable to the one computed in Equation (2.8).

The GCI for γ_P is computed using the GPQ shown in Equation (2.7). For the given data set, this GPQ is

$$\text{GPQ}(\gamma_P) = \max\left[0, \frac{(p-1)s_P^2}{rW_1} - \frac{p(r-1)s_E^2}{rW_2}\right]$$
$$= \max\left[0, \frac{9(94.47)}{2W_1} - \frac{10(1)(0.50)}{2W_2}\right], \tag{2.9}$$

where W_1 and W_2 are independent chi-squared random variables with 9 and 10 degrees of freedom, respectively. We simulate $N = 100{,}000$ values of Equation (2.9) by simulating 100,000 values each of W_1 and W_2. The 100,000 GPQ values are sorted from least to greatest. The lower bound for a 95% confidence interval is the value in position $100{,}000 \times (0.05/2) = 2{,}500$ of the ordered set of simulated GPQ values, and the upper bound is the value in position $100{,}000 \times (1 - 0.05/2) = 97{,}500$. For our simulated values of W_1 and W_2, the 95% generalized interval for γ_P is from $L = 22.0$ to $U = 157$. As is often the case, this interval is very similar to the closed-form intervals computed for this example.

2.5.3 Interval for γ_M

The point estimate for γ_M is $S_E^2 = 0.50$. The exact 95% confidence interval from Equation (2.5) is

$$L = (1 - G_2)S_E^2 = (1 - 0.5118)(0.50) = 0.244$$

and

$$U = (1 + H_2)S_E^2 = (1 + 2.080)(0.50) = 1.54. \tag{2.10}$$

2.5.4 Interval for γ_R

The point estimate for γ_R is

$$\widehat{\gamma}_R = \frac{S_P^2/S_E^2 - 1}{r}$$
$$= \frac{94.47/0.50 - 1}{2}$$
$$= 94.0.$$

The 95% confidence interval for γ_R based on Equation (2.6) is

$$\begin{aligned} L &= \frac{S_P^2}{rS_E^2F_1} - \frac{1}{r} \\ &= \frac{94.47}{2(0.50)(3.7790)} - \frac{1}{2} \\ &= 24.5 \end{aligned}$$

and

$$\begin{aligned} U &= \frac{S_P^2}{rS_E^2F_2} - \frac{1}{r} \\ &= \frac{94.47}{2(0.50)(0.2523)} - \frac{1}{2} \\ &= 374. \end{aligned} \tag{2.11}$$

2.5.5 Interval for PTR

As defined in Equation (1.2), the PTR is

$$\text{PTR} = \frac{k\sqrt{\gamma_M}}{USL - LSL}.$$

For this example, the specification limits are $LSL = 18$ and $USL = 58$. Using $k = 6$, a point estimate for PTR is

$$\frac{k\sqrt{\widehat{\gamma}_M}}{USL - LSL} = \frac{6\sqrt{0.50}}{58 - 18} = 0.106.$$

A 95% confidence interval for PTR based on the computed confidence interval for γ_M in Equation (2.10) is

$$L = \frac{6\sqrt{0.244}}{58 - 18} = 0.074$$

and

$$U = \frac{6\sqrt{1.54}}{58 - 18} = 0.186.$$

2.5.6 Interval for SNR

As defined in Equation (1.3), the SNR is expressed as

$$\text{SNR} = \sqrt{2\gamma_R}.$$

A point estimate for SNR is

$$\sqrt{2\widehat{\gamma}_R} = \sqrt{2(94.0)} = 13.7.$$

A 95% confidence interval for SNR based on the computed confidence interval for γ_R in Equation (2.11) is

$$L = \sqrt{2(24.5)} = 7.00$$

and

$$U = \sqrt{2(374)} = 27.3.$$

2.5.7 Interval for C_p

Using the relation in Equation (1.5), a point estimate for C_p is

$$\frac{USL - LSL}{6\sqrt{\widehat{\gamma}_P}} = \frac{58 - 18}{6\sqrt{47.0}} = 0.972.$$

The 95% confidence interval for C_p based on the computed MLS interval in Equation (2.8) is

$$L = \frac{58 - 18}{6\sqrt{157}} = 0.532$$

and

$$U = \frac{58 - 18}{6\sqrt{22.1}} = 1.42.$$

2.5.8 Intervals for Misclassification Rates

Table 2.9 reports the 95% confidence intervals for the misclassification rates using the algorithm described in Section 2.4 with $N = 100{,}000$ and $\epsilon = 0.001$. All values in the intervals for δ_{index} and β_{index} are less than one, which suggests the system is better than a chance process.

Table 2.9. *95% confidence intervals for misclassification rates. (Bounds for δ, δ_c, β, and β_c have been multiplied by 10^6.)*

Parameter	Lower bound	Upper bound
δ	42	7,433
δ_c	42	8,654
δ_{index}	0.046	0.459
β	20	6,406
β_c	36,330	197,551
β_{index}	0.042	0.198

Table 2.10. *95% confidence intervals for example. (See Section* 1.8 *for a description of computer programs to perform these computations.)*

Parameter	Lower bound	Upper bound
μ_Y	29.9	39.7
γ_P	22.1	157
γ_M	0.244	1.54
γ_R	24.5	374
PTR	0.074	0.186
SNR	7.00	27.3
C_p	0.532	1.42

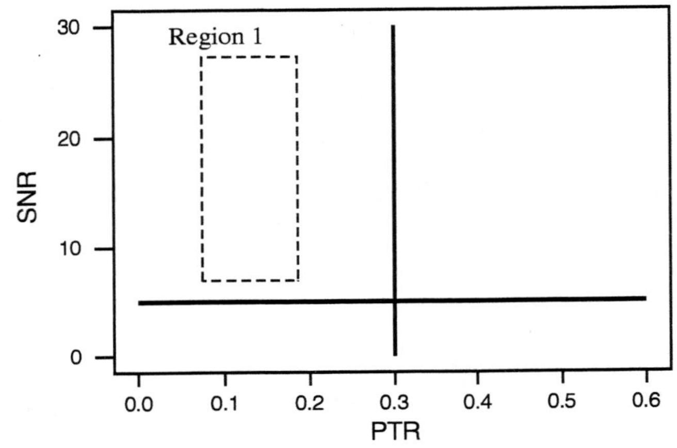

Figure 2.1. *R&R graph for one-factor example.*

2.5.9 Conclusions

Table 2.10 reports the computed bounds for the gauge R&R parameters in this example. Results for γ_P and C_p are based on the MLS interval.

This particular measurement system appears to be capable of monitoring the manufacturing process. Figure 2.1 shows the R&R graph described in Section 1.5, where $k = 6$ and the dark lines are placed at PTR $= 0.3$ and SNR $= 5$. The dashed lines represent the intersection of the confidence intervals for PTR and SNR. This intersection is contained in Region 1 and implies the measurement system satisfies both criteria. Additionally, the misclassification indexes in Table 2.9 are less than one, and this suggests the system is able to discriminate between good and bad parts.

2.6 Summary

In this chapter we presented a gauge R&R study in which a single operator measures each part in a random sample a fixed number of times. The model used to describe this process is the one-factor random design. Chapter 3 considers the situation in which a sample of operators is used rather than a single operator to make the replicated measurement of parts. This is the classical design layout for a gauge R&R study. The statistical model used to represent this process is the two-factor random model with interaction.

Chapter 3

Balanced Two-Factor Crossed Random Models with Interaction

3.1 Introduction

The two-factor crossed design with interaction is the classical gauge R&R model. Typically, the two factors are referred to as "parts" and "operators." In this chapter we consider this design for balanced experiments where both factors are random. Extensions to situations where either (i) one factor is fixed or (ii) the design is unbalanced are considered in Chapters 6 and 7, respectively. The two-factor model with no interaction is presented in Chapter 5.

We begin with an example that will be used throughout this chapter. Table 3.1 reports a partial listing of a data set based on the experiment described by Houf and Berman [32]. We have modified the experiment to increase the number of operators and parts used by Houf and Berman. The response variable is the thermal performance of a module measured in C° per watt. Each response has been multiplied by 100 for convenience of scale. The data represent measurements of 20 parts recorded by six operators. Each part is measured two times by each operator. In this chapter, it is assumed that parts and operators are selected at random from larger populations. We now consider a model to represent the two-factor data shown in Table 3.1.

3.2 The Model

The balanced two-factor crossed random model with interaction is

$$Y_{ijk} = \mu_Y + P_i + O_j + (PO)_{ij} + E_{ijk}, \qquad (3.1)$$
$$i = 1, \ldots, p, \quad j = 1, \ldots, o, \quad k = 1, \ldots, r,$$

where μ_Y is a constant and P_i, O_j, $(PO)_{ij}$, E_{ijk} are jointly independent normal random variables with means of zero and variances σ_P^2, σ_O^2, σ_{PO}^2, and σ_E^2, respectively. The ANOVA for model (3.1) is shown in Table 3.2, and the definitions for the mean squares and means are shown in Table 3.3. Table 3.4 reports distributional properties based on the assumptions in model (3.1).

Table 3.1. *Example data for balanced two-factor random model.*

Part	Operator 1	Operator 2	...	Operator 6
1	45, 44	43, 44	...	46, 46
2	21, 23	20, 22	...	21, 21
$\vdots$	$\vdots$	$\vdots$	...	$\vdots$
19	32, 33	32, 35	...	33, 33
20	29, 31	31, 30	...	31, 29

Table 3.2. *ANOVA for model* (3.1).

Source of variation	Degrees of freedom	Mean square	Expected mean square
Parts (P)	$p-1$	S_P^2	$\theta_P = \sigma_E^2 + r\sigma_{PO}^2 + or\sigma_P^2$
Operators (O)	$o-1$	S_O^2	$\theta_O = \sigma_E^2 + r\sigma_{PO}^2 + pr\sigma_O^2$
P×O	$(p-1)(o-1)$	S_{PO}^2	$\theta_{PO} = \sigma_E^2 + r\sigma_{PO}^2$
Replicates	$po(r-1)$	S_E^2	$\theta_E = \sigma_E^2$

Table 3.3. *Mean squares and means for model* (3.1).

Statistic	Definition
S_P^2	$\dfrac{or\Sigma_i(\overline{Y}_{i**} - \overline{Y}_{***})^2}{p-1}$
S_O^2	$\dfrac{pr\Sigma_j(\overline{Y}_{*j*} - \overline{Y}_{***})^2}{o-1}$
S_{PO}^2	$\dfrac{r\Sigma_i\Sigma_j(\overline{Y}_{ij*} - \overline{Y}_{i**} - \overline{Y}_{*j*} + \overline{Y}_{***})^2}{(p-1)(o-1)}$
S_E^2	$\dfrac{\Sigma_i\Sigma_j\Sigma_k(Y_{ijk} - \overline{Y}_{ij*})^2}{po(r-1)}$
$\overline{Y}_{i**}$	$\dfrac{\Sigma_j\Sigma_k Y_{ijk}}{or}$
$\overline{Y}_{*j*}$	$\dfrac{\Sigma_i\Sigma_k Y_{ijk}}{pr}$
$\overline{Y}_{ij*}$	$\dfrac{\Sigma_k Y_{ijk}}{r}$
$\overline{Y}_{***}$	$\dfrac{\Sigma_i\Sigma_j\Sigma_k Y_{ijk}}{por}$

Table 3.4. *Distributional results for model* (3.1).

Result	
1	$\overline{Y}_{***}$, S_P^2, S_O^2, S_{PO}^2, and S_E^2 are jointly independent.
2	$(p-1)S_P^2/\theta_P$ is a chi-squared random variable with $p-1$ degrees of freedom.
3	$(o-1)S_O^2/\theta_O$ is a chi-squared random variable with $o-1$ degrees of freedom.
4	$(p-1)(o-1)S_{PO}^2/\theta_{PO}$ is a chi-squared random variable with $(p-1)(o-1)$ degrees of freedom.
5	$po(r-1)S_E^2/\theta_E$ is a chi-squared random variable with $po(r-1)$ degrees of freedom.
6	$\overline{Y}_{***}$ is a normal random variable with mean μ_Y and variance $\dfrac{\theta_P+\theta_O-\theta_{PO}}{por}$.

Table 3.5. *Covariance structure for model* (3.1).

Condition	Covariance$(Y_{ijk}, Y_{i'j'k'})$
$i=i', j=j', k\neq k'$ (same part and same operator)	$\sigma_P^2+\sigma_O^2+\sigma_{PO}^2$
$i=i', j\neq j'$ (same part with different operators)	σ_P^2
$i\neq i', j=j'$ (same operator with different parts)	σ_O^2
$i\neq i', j\neq j'$ (different parts and operators)	0

The term (PO_{ij}) in model (3.1) is commonly referred to as an interaction effect. Interaction is included in a random effects model to provide more flexibility in modeling the covariance structure of the observations. Model (3.1) implies the covariance structure shown in Table 3.5. Notice that if $\sigma_{PO}^2 > 0$, the covariance between repeated measurements on the same part by the same operator is greater than the sum of the individual variances σ_P^2 and σ_O^2. A statistical test to determine if there is evidence that $\sigma_{PO}^2 > 0$ is described in Section 3.8.1.

3.3 MLS Intervals, Gauge R&R Parameters

Table 3.6 reports gauge R&R parameters and point estimators for model (3.1). Variation of the measurement system is attributed to all sources of variation except parts. Thus, $\gamma_M = \sigma_O^2+\sigma_{PO}^2+\sigma_E^2$. The estimators for μ_Y, γ_P, and γ_M are MVU estimators.

Table 3.6. *Gauge R&R parameters and point estimators for model* (3.1).

Gauge R&R notation	Model (3.1) representation	Point estimator
μ_Y	μ_Y	$\widehat{\mu}_Y = \overline{Y}_{***}$
γ_P	σ_P^2	$\widehat{\gamma}_P = \dfrac{S_P^2 - S_{PO}^2}{or}$
γ_M	$\sigma_O^2 + \sigma_{PO}^2 + \sigma_E^2$	$\widehat{\gamma}_M = \dfrac{S_O^2 + (p-1)S_{PO}^2 + p(r-1)S_E^2}{pr}$
γ_R	$\dfrac{\sigma_P^2}{\sigma_O^2 + \sigma_{PO}^2 + \sigma_E^2}$	$\widehat{\gamma}_R = \dfrac{\widehat{\gamma}_P}{\widehat{\gamma}_M}$

Table 3.7. *Constants used in confidence intervals for model* (3.1). *Values are for* $\alpha = 0.05$, $p = 20$, $o = 6$, *and* $r = 2$.

Constant	Definition	Value
G_1	$1 - F_{\alpha/2:\infty,p-1}$	0.4217
G_2	$1 - F_{\alpha/2:\infty,o-1}$	0.6104
G_3	$1 - F_{\alpha/2:\infty,(p-1)(o-1)}$	0.2330
G_4	$1 - F_{\alpha/2:\infty,po(r-1)}$	0.2116
H_1	$F_{1-\alpha/2:\infty,p-1} - 1$	1.133
H_2	$F_{1-\alpha/2:\infty,o-1} - 1$	5.015
H_3	$F_{1-\alpha/2:\infty,(p-1)(o-1)} - 1$	0.3586
H_4	$F_{1-\alpha/2:\infty,po(r-1)} - 1$	0.3104
F_1	$F_{1-\alpha/2:p-1,(p-1)(o-1)}$	1.876
F_2	$F_{\alpha/2:p-1,(p-1)(o-1)}$	0.4503
F_3	$F_{1-\alpha/2:p-1,o-1}$	6.344
F_4	$F_{\alpha/2:p-1,o-1}$	0.3001
G_{13}	$\dfrac{(F_1-1)^2 - G_1^2F_1^2 - H_3^2}{F_1}$	0.006771
H_{13}	$\dfrac{(1-F_2)^2 - H_1^2F_2^2 - G_3^2}{F_2}$	−0.0276

We now present MLS confidence intervals for the parameters defined in Table 3.6. Generalized confidence intervals for these parameters are presented in Section 3.4. Table 3.7 reports constants required to compute the MLS confidence intervals, where $F_{\alpha:df1,df2}$ represents the F-value with $df1$ and $df2$ degrees of freedom and area α to the left. The particular values reported in this table provide a two-sided interval with a 95% confidence coefficient ($\alpha = 0.05$) for a design with $p = 20$ parts, $o = 6$ operators, and $r = 2$ replicates.

3.3.1 Interval for μ_Y

Paark and Burdick [55] studied several methods for constructing confidence intervals on μ_Y in model (3.1). A relatively easy interval to compute is based on the Satterthwaite [57, 58] approximation. However, the confidence coefficient for this interval can be less than the stated level, and this situation is most likely to occur when the difference between p and o is great. Since p is often much greater than o in a gauge R&R study, an alternative confidence interval is recommended. A computationally simple $100(1-\alpha)\%$ interval recommended by Milliken and Johnson [50, p. 281] is

$$L = \overline{Y}_{***} - C\sqrt{\frac{K}{por}}$$

and

$$U = \overline{Y}_{***} + C\sqrt{\frac{K}{por}}, \tag{3.2}$$

where

$$K = S_P^2 + S_O^2 - S_{PO}^2$$

and

$$C = \frac{S_P^2\sqrt{F_{1-\alpha:1,p-1}} + S_O^2\sqrt{F_{1-\alpha:1,o-1}} - S_{PO}^2\sqrt{F_{1-\alpha:1,(p-1)(o-1)}}}{K}.$$

If $K < 0$, then replace K with S_{PO}^2 and C with $\sqrt{F_{1-\alpha:1,(p-1)(o-1)}}$. Interval (3.2) maintains the stated confidence level and provided relatively short intervals in the Paark and Burdick study.

3.3.2 Interval for γ_P

We begin by writing γ_P in terms of expected mean squares as $\gamma_P = \sigma_P^2 = (\theta_P - \theta_{PO})/(or)$. Since this parameter involves the difference of two mean squares, we use the MLS method proposed by Ting et al. [64]. Based on this method, the bounds of an approximate $100(1-\alpha)\%$ confidence interval for γ_P are

$$L = \widehat{\gamma}_P - \frac{\sqrt{V_{LP}}}{or}$$

and

$$U = \widehat{\gamma}_P + \frac{\sqrt{V_{UP}}}{or}, \tag{3.3}$$

where

$$V_{LP} = G_1^2 S_P^4 + H_3^2 S_{PO}^4 + G_{13} S_P^2 S_{PO}^2,$$

$$V_{UP} = H_1^2 S_P^4 + G_3^2 S_{PO}^4 + H_{13} S_P^2 S_{PO}^2,$$

$\widehat{\gamma}_P$ is defined in Table 3.6, and G_1, G_3, H_1, H_3, G_{13}, and H_{13} are defined in Table 3.7. Negative bounds are increased to zero.

The Tukey–Williams method described in Section 2.3.2 can also be used to compute an interval for γ_P. This interval will provide similar results to the interval in Equation (3.3).

3.3.3 Interval for γ_M

The variability of the measurement process is $\gamma_M = \sigma_O^2 + \sigma_{PO}^2 + \sigma_E^2 = [\theta_O + (p-1)\theta_{PO} + p(r-1)\theta_E]/(pr)$. Since γ_M is a sum of expected mean squares, we apply the MLS method proposed by Graybill and Wang [25]. The resulting bounds of an approximate $100(1-\alpha)\%$ confidence interval for γ_M are

$$L = \widehat{\gamma}_M - \frac{\sqrt{V_{LM}}}{pr}$$

and

$$U = \widehat{\gamma}_M + \frac{\sqrt{V_{UM}}}{pr}, \tag{3.4}$$

where

$$V_{LM} = G_2^2 S_O^4 + G_3^2 (p-1)^2 S_{PO}^4 + G_4^2 p^2 (r-1)^2 S_E^4,$$
$$V_{UM} = H_2^2 S_O^4 + H_3^2 (p-1)^2 S_{PO}^4 + H_4^2 p^2 (r-1)^2 S_E^4,$$

$\widehat{\gamma}_M$ is defined in Table 3.6, and G_2, G_3, G_4, H_2, H_3, and H_4 are defined in Table 3.7.

3.3.4 Interval for γ_R

Leiva and Graybill [42] proposed a confidence interval for γ_R in model (3.1). Chiang [15] demonstrated that this interval is preferable to the one recommended by Burdick and Larsen [11]. This approximate $100(1-\alpha)\%$ confidence interval is

$$L = \frac{p(1-G_1)(S_P^2 - F_1 S_{PO}^2)}{po(r-1)S_E^2 + o(1-G_1)F_3 S_O^2 + o(p-1)S_{PO}^2}$$

and

$$U = \frac{p(1+H_1)(S_P^2 - F_2 S_{PO}^2)}{po(r-1)S_E^2 + o(1+H_1)F_4 S_O^2 + o(p-1)S_{PO}^2} \tag{3.5}$$

where G_1, H_1, F_1, F_2, F_3, and F_4 are defined in Table 3.7. Negative bounds are increased to zero.

3.4 Generalized Intervals, Gauge R&R Parameters

Since there are no exact intervals for any of the parameters in Table 3.6, we define GPQs for all the parameters. Table 3.8 reports these GPQs where Z is a standard normal random variable and W_1, W_2, W_3, and W_4 are jointly independent chi-squared random variables that are independent of Z with degrees of freedom $p-1$, $o-1$, $(p-1)(o-1)$, and $po(r-1)$,

Table 3.8. *GPQs for gauge R&R parameters in model* (3.1).

Parameter	GPQ
μ_Y	$\overline{y}_{***} - Z\sqrt{\max\left[\epsilon, \frac{(p-1)s_P^2}{por\,W_1} + \frac{(o-1)s_O^2}{por\,W_2} - \frac{(p-1)(o-1)s_{PO}^2}{por\,W_3}\right]}$
γ_P	$\max\left[0, \frac{(p-1)s_P^2}{or\,W_1} - \frac{(p-1)(o-1)s_{PO}^2}{or\,W_3}\right]$
γ_M	$\frac{(o-1)s_O^2}{pr\,W_2} + \frac{(p-1)^2(o-1)s_{PO}^2}{pr\,W_3} + \frac{po(r-1)^2 s_E^2}{r\,W_4}$
γ_R	$\frac{\text{GPQ}(\gamma_P)}{\text{GPQ}(\gamma_M)}$

respectively. The terms s_P^2, s_O^2, s_{PO}^2, and s_E^2 are the realized values of the mean squares in the particular data set. The symbol ϵ is a small positive quantity needed to keep the term under the square root of GPQ(μ_Y) positive. The GPQs for γ_P and γ_M were recommended by Hamada and Weerahandi [27] and correspond to the tailored variables recommended by Chiang [14] in Table 4 of his paper.

Consistent with our approach in Chapter 2, we construct the generalized intervals in the following manner:

1. Compute $\overline{Y}_{***}$, S_P^2, S_O^2, S_{PO}^2, and S_E^2 for the collected data and denote the realized values as $\overline{y}_{***}$, s_P^2, s_O^2, s_{PO}^2, and s_E^2, respectively.
2. Simulate N values of the appropriate GPQ shown in Table 3.8 by simulating N independent values each of W_1, W_2, W_3, W_4, and Z.
3. Order the N GPQ values in step 2 from least to greatest.
4. Define the lower bound for a $100(1-\alpha)\%$ interval as the value in position $N \times (\alpha/2)$ of the ordered set in step 3. Define the upper bound as the value in position $N \times (1-\alpha/2)$ of this same ordered set.

Generalized confidence intervals for the misclassification rates are computed using the GPQ values shown in Table 3.9. To demonstrate, the following algorithm is used to compute the confidence interval for δ:

1. Compute $\overline{Y}_{***}$, S_P^2, S_O^2, S_{PO}^2, and S_E^2 for the collected data and denote the realized values as $\overline{y}_{***}$, s_P^2, s_O^2, s_{PO}^2, and s_E^2, respectively.
2. Simulate N values of each GPQ defined in Table 3.9. This is done by simulating N independent values each of W_1, W_2, W_3, W_4, and Z.
3. Compute N values of δ using Equation (1.9) by replacing μ_Y, μ_P, $\gamma_P + \gamma_M$, and γ_P with the GPQs formed in step 2.
4. Order the N values obtained in step 3 from least to greatest.
5. Define the lower bound for a $100(1-\alpha)\%$ interval as the value in position $N \times (\alpha/2)$ of the ordered set in step 4. Define the upper bound as the value in position $N \times (1-\alpha/2)$ of this same ordered set.

Table 3.9. *GPQs for misclassification rates in model* (3.1).

Parameter	GPQ
$\mu_Y = \mu_P$	$\bar{y}_{***} - Z\sqrt{\max\left[\epsilon, \frac{(p-1)s_P^2}{por\,W_1} + \frac{(o-1)s_O^2}{por\,W_2} - \frac{(p-1)(o-1)s_{PO}^2}{por\,W_3}\right]}$
$\gamma_P + \gamma_M$	$\frac{(p-1)s_P^2}{or\,W_1} + \frac{(o-1)s_O^2}{pr\,W_2} + \frac{(po-p-o)(p-1)(o-1)s_{PO}^2}{por\,W_3} + \frac{po(r-1)^2 s_E^2}{r\,W_4}$
γ_P	$\max\left[\epsilon, \frac{(p-1)s_P^2}{or\,W_1} - \frac{(p-1)(o-1)s_{PO}^2}{or\,W_3}\right]$

Table 3.10. *ANOVA for two-factor example.*

Source of variation	Degrees of freedom	Mean square
Parts (P)	19	591.5
Operators (O)	5	13.59
P×O	95	2.060
Replicates	120	0.7333

Complete details for computing intervals for the misclassification rates are found in the paper by Burdick et al. [13]. This process and all the formulas presented in this section are demonstrated in the following numerical example.

3.5 Numerical Example

We now perform the analysis for the data in Table 3.1 using the ANOVA shown in Table 3.10. To construct two-sided 95% confidence intervals, we use the constants shown in Table 3.7 with $\alpha = 0.05$, $p = 20$ parts, $o = 6$ operators, and $r = 2$ replicates. The mean of all the observations is 36.9. The specification limits needed to define PTR and the misclassification rates are $LSL = 18$ and $USL = 58$. The value used for ϵ in the GCI computations is 0.001.

3.5.1 Intervals for μ_Y

The confidence interval for μ_Y is shown in Equation (3.2). For our data $K = 603.0$ and $C = 2.104$ using $\alpha = 0.05$ with $F_{0.95:1,19} = 4.3807$, $F_{0.95:1,5} = 6.6079$, and

$F_{0.95:1,95} = 3.9412$. Hence the 95% confidence interval is

$$L = \overline{Y}_{***} - C\sqrt{\frac{K}{por}}$$
$$= 36.9 - 2.104\sqrt{\frac{603.0}{20(6)(2)}}$$
$$= 33.5$$

and

$$U = \overline{Y}_{***} + C\sqrt{\frac{K}{por}}$$
$$= 36.9 + 2.104\sqrt{\frac{603.0}{20(6)(2)}}$$
$$= 40.2. \tag{3.6}$$

The generalized confidence interval for μ_Y is computed using Table 3.8 with

$$\text{GPQ}(\mu_Y) = \overline{y}_{***} - Z\sqrt{\max\left[\epsilon, \frac{(p-1)s_P^2}{por\,W_1} + \frac{(o-1)s_O^2}{por\,W_2} - \frac{(p-1)(o-1)s_{PO}^2}{por\,W_3}\right]}$$
$$= 36.9 - Z\sqrt{\max\left[0.001, \frac{19(591.5)}{240W_1} + \frac{5(13.59)}{240W_2} - \frac{95(2.060)}{240W_3}\right]}, \tag{3.7}$$

where Z is a standard normal random variable and W_1, W_2, and W_3 are jointly independent chi-squared random variables that are independent of Z with degrees of freedom 19, 5, and 95, respectively. We now simulate N = 100,000 values of Equation (3.7) and sort them from least to greatest. The lower bound for a 95% generalized confidence interval is the value in position 100,000 × (0.05/2) = 2,500 of the ordered set of simulated GPQs. The upper bound is the value in position 100,000 × (1 − 0.05/2) = 97,500 of this same ordered set. For our set of simulated values, the 95% generalized interval for μ_Y is from $L = 33.5$ to $U = 40.2$. This interval is identical to the one computed in Equation (3.6).

3.5.2 Intervals for γ_P

The point estimate for γ_P is

$$\widehat{\gamma}_P = \frac{S_P^2 - S_{PO}^2}{or}$$
$$= \frac{591.5 - 2.060}{12}$$
$$= 49.1.$$

The lower and upper confidence bounds are shown in Equation (3.3). In this example,

$$\begin{aligned}V_{LP} &= G_1^2 S_P^4 + H_3^2 S_{PO}^4 + G_{13} S_P^2 S_{PO}^2 \\ &= (0.4217)^2(591.5)^2 + (0.3586)^2(2.060)^2 + (0.006771)(591.5)(2.060) \\ &= 6.22 \times 10^4\end{aligned}$$

and

$$\begin{aligned}V_{UP} &= H_1^2 S_P^4 + G_3^2 S_{PO}^4 + H_{13} S_P^2 S_{PO}^2 \\ &= (1.133)^2(591.5)^2 + (0.2330)^2(2.060)^2 + (-0.0276)(591.5)(2.060) \\ &= 4.49 \times 10^5.\end{aligned}$$

Substituting this information into Equation (3.3) yields the 95% confidence interval

$$L = 49.1 - \frac{\sqrt{6.22 \times 10^4}}{12} = 28.3$$

and

$$U = 49.1 + \frac{\sqrt{4.49 \times 10^5}}{12} = 105. \tag{3.8}$$

The generalized confidence interval for γ_P is computed using the GPQ for γ_P shown in Table 3.8. For the given data set, this GPQ is

$$\begin{aligned}\text{GPQ}(\gamma_P) &= \max\left[0, \frac{(p-1)s_P^2}{or W_1} - \frac{(p-1)(o-1)s_{PO}^2}{or W_3}\right] \\ &= \max\left[0, \frac{19(591.5)}{12W_1} - \frac{95(2.060)}{12W_3}\right],\end{aligned}$$

where W_1 and W_3 are independent chi-squared random variables with 19 and 95 degrees of freedom, respectively. The computed 95% generalized interval for γ_P based on our simulated values of W_1 and W_3 is from $L = 28.4$ to $U = 105$.

3.5.3 Intervals for γ_M

The point estimate for γ_M is

$$\begin{aligned}\widehat{\gamma}_M &= \frac{S_O^2 + (p-1)S_{PO}^2 + p(r-1)S_E^2}{pr} \\ &= \frac{13.59 + 19(2.060) + 20(0.7333)}{40} \\ &= 1.685.\end{aligned}$$

The lower and upper confidence bounds are shown in Equation (3.4). In this example

$$\begin{aligned}V_{LM} &= G_2^2 S_O^4 + G_3^2(p-1)^2 S_{PO}^4 + G_4^2 p^2 (r-1)^2 S_E^4 \\ &= (0.6104)^2(13.59)^2 + (0.2330)^2(19)^2(2.060)^2 + (0.2116)^2(20)^2(0.7333)^2 \\ &= 161.6\end{aligned}$$

and

$$\begin{aligned} V_{UM} &= H_2^2 S_O^4 + H_3^2 (p-1)^2 S_{PO}^4 + H_4^2 p^2 (r-1)^2 S_E^4 \\ &= (5.015)^2(13.59)^2 + (0.3586)^2(19)^2(2.060)^2 + (0.3104)^2(20)^2(0.7333)^2 \\ &= 4{,}862. \end{aligned}$$

This yields the 95% confidence interval

$$L = 1.685 - \frac{\sqrt{161.6}}{40} = 1.37$$

and

$$U = 1.685 + \frac{\sqrt{4,862}}{40} = 3.43. \tag{3.9}$$

The GPQ for γ_M shown in Table 3.8 is

$$\begin{aligned} \text{GPQ}(\gamma_M) &= \frac{(o-1)s_O^2}{prW_2} + \frac{(p-1)^2(o-1)s_{PO}^2}{prW_3} + \frac{po(r-1)^2 s_E^2}{rW_4} \\ &= \frac{5(13.59)}{40W_2} + \frac{(19)^2(5)(2.060)}{40W_3} + \frac{120(0.7333)}{2W_4}, \end{aligned}$$

where W_2, W_3, and W_4 are independent chi-squared random variables with 5, 95, and 120 degrees of freedom, respectively. Our computed 95% generalized interval for γ_M is from $L = 1.36$ to $U = 3.42$.

3.5.4 Interval for γ_R

The point estimate for γ_R is

$$\begin{aligned} \widehat{\gamma}_R &= \frac{\widehat{\gamma}_P}{\widehat{\gamma}_M} \\ &= \frac{49.1}{1.685} \\ &= 29.1. \end{aligned}$$

Using Equation (3.5) to compute a 95% confidence interval, we obtain

$$\begin{aligned} L &= \frac{p(1-G_1)(S_P^2 - F_1 S_{PO}^2)}{po(r-1)S_E^2 + o(1-G_1)F_3 S_O^2 + o(p-1)S_{PO}^2} \\ &= \frac{20(1-0.4217)[591.5 - 1.876(2.060)]}{120(0.7333) + 6(1-0.4217)(6.344)(13.59) + 114(2.060)} \\ &= 10.9 \end{aligned}$$

and

$$U = \frac{p(1 + H_1)(S_P^2 - F_2 S_{PO}^2)}{po(r-1)S_E^2 + o(1+H_1)F_4 S_O^2 + o(p-1)S_{PO}^2}$$
$$= \frac{20(1 + 1.133)[591.5 - 0.4503(2.060)]}{120(0.7333) + 6(1 + 1.133)(0.3001)(13.59) + 114(2.060)}$$
$$= 67.2. \tag{3.10}$$

The 95% generalized confidence interval formed using the GPQ for γ_R shown in Table 3.8 is from $L = 12.0$ to $U = 61.5$.

3.5.5 Interval for PTR

Using Equation (1.2) with $k = 6$, a point estimate for PTR is

$$\frac{k\sqrt{\widehat{\gamma}_M}}{USL - LSL} = \frac{6\sqrt{1.685}}{58 - 18} = 0.195.$$

A 95% confidence interval for PTR based on the bounds computed in Equation (3.9) is

$$L = \frac{6\sqrt{1.37}}{58 - 18} = 0.175$$

and

$$U = \frac{6\sqrt{3.43}}{58 - 18} = 0.278.$$

3.5.6 Interval for SNR

Based on Equation (1.3), the point estimate for SNR is

$$\sqrt{2\widehat{\gamma}_R} = \sqrt{2(29.1)} = 7.63.$$

A 95% confidence interval for SNR based on the bounds for γ_R calculated in Equation (3.10) is

$$L = \sqrt{2(10.9)} = 4.67$$

and

$$U = \sqrt{2(67.2)} = 11.6.$$

3.5.7 Interval for C_p

Using the relation in Equation (1.5) and the point estimate for γ_P, a point estimate for C_p is

$$\frac{58 - 18}{6\sqrt{49.1}} = 0.951.$$

Table 3.11. *95% confidence intervals for misclassification rates. (Bounds for δ, δ_c, β, and β_c have been multiplied by 10^6.)*

Parameter	Lower bound	Upper bound
δ	199	8,594
δ_c	199	9,135
δ_{index}	0.141	0.854
β	63	6,138
β_c	98,746	248,703
β_{index}	0.105	0.249

The 95% confidence interval for C_p based on the computed interval in Equation (3.8) is

$$L = \frac{58 - 18}{6\sqrt{105}} = 0.651$$

and

$$U = \frac{58 - 18}{6\sqrt{28.3}} = 1.25.$$

3.5.8 Intervals for Misclassification Rates

Table 3.11 reports 95% confidence intervals for the misclassification rates. These intervals were computed using the algorithm described in Section 3.4 with $N = 100{,}000$ and $\epsilon = 0.001$. Since the upper bounds for δ_{index} and β_{index} are less than one, there is evidence that the system is better than what one would expect by chance.

3.5.9 Conclusions

Table 3.12 reports the 95% MLS intervals and GCIs for all parameters in the example. As shown in the table, the results for the two methods are comparable. The only practical difference is that the GCI for γ_R (and hence SNR) is shorter than the MLS interval.

The results suggest the measurement system is capable. Figure 3.1 shows the R&R graph described in Section 1.5 with $k = 6$ and the desired criteria PTR ≤ 0.3 and SNR ≥ 5. The dashed rectangle formed by the intersection of the confidence intervals for PTR and SNR lies mostly in Region 1. Additionally, the misclassification rates appear to be at an acceptable level.

3.6 Comparing Repeatability and Reproducibility

As noted in Chapter 1, components of the measurement system variation (γ_M) have traditionally been labeled as repeatability and reproducibility. Repeatability represents the gauge variability when used to measure the same part. In the context of model (3.1) repeatability is defined as the variance component σ_E^2. Reproducibility refers to the variability arising from different operators and is defined as the sum of the variance components $\sigma_O^2 + \sigma_{PO}^2$.

Table 3.12. *Comparison of 95% MLS and generalized confidence intervals. (See Section 1.8 for a description of computer programs to perform these computations.)*

Parameter	MLS	GCI
μ_Y	$L = 33.5$	$L = 33.5$
	$U = 40.2$	$U = 40.2$
γ_P	$L = 28.3$	$L = 28.4$
	$U = 105$	$U = 105$
γ_M	$L = 1.37$	$L = 1.36$
	$U = 3.43$	$U = 3.42$
γ_R	$L = 10.9$	$L = 12.0$
	$U = 67.2$	$U = 61.5$
PTR	$L = 0.175$	$L = 0.175$
	$U = 0.278$	$U = 0.277$
SNR	$L = 4.67$	$L = 4.91$
	$U = 11.6$	$U = 11.1$
C_p	$L = 0.651$	$L = 0.652$
	$U = 1.25$	$U = 1.25$

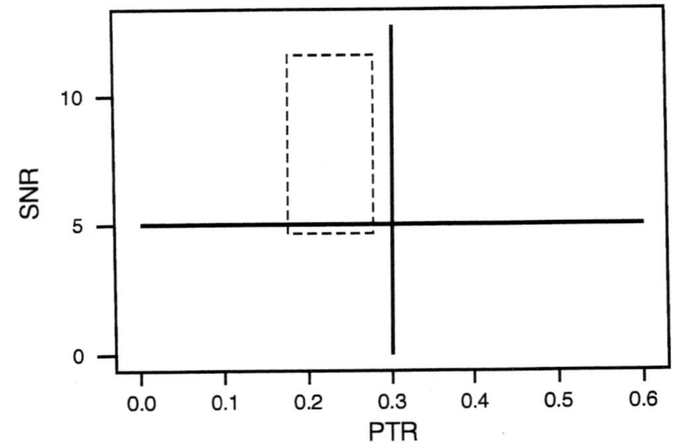

Figure 3.1. *R&R graph for two-factor example.*

By comparing the relative magnitudes of repeatability and reproducibility, one can often determine the best approach for improving a measurement system. In many cases it is easier and less expensive to reduce reproducibility. For example, it might be relatively easy to improve performance of human operators by implementing a short training program. In contrast, reduction of repeatability often requires technical modifications to the gauge. Thus, it is useful to know what percentage of γ_M is attributed to each component. To do this, we require a confidence interval for the ratio $\rho_{repeat} = \sigma_E^2/\gamma_M$. This parameter represents the proportion of measurement system variation due to repeatability. The proportion due to reproducibility is $1 - \rho_{repeat}$.

To begin, we consider the ratio

$$\begin{aligned}\eta &= \frac{\sigma_O^2 + \sigma_{PO}^2}{\sigma_E^2} \\ &= \frac{\theta_O + (p-1)\theta_{PO} - p\theta_E}{pr\theta_E} \\ &= \frac{\theta_O + (p-1)\theta_{PO}}{pr\theta_E} - \frac{1}{r}.\end{aligned}$$

A confidence interval for η can be constructed and then converted to an interval for ρ_{repeat} using the relation

$$\rho_{repeat} = \frac{1}{\eta + 1}. \tag{3.11}$$

Based on the results of Lu, Graybill, and Burdick [45], a $100(1-\alpha)\%$ interval for η is

$$L = \frac{1}{prS_E^2}\left[\left(1 - \frac{2}{po(r-1)}\right)\left(S_O^2 + (p-1)S_{PO}^2\right) - \sqrt{V_L}\right] - \frac{1}{r}$$

and

$$U = \frac{1}{prS_E^2}\left[\left(1 - \frac{2}{po(r-1)}\right)\left(S_O^2 + (p-1)S_{PO}^2\right) + \sqrt{V_U}\right] - \frac{1}{r}, \tag{3.12}$$

where

$$V_L = a_L S_O^4 + b_L (p-1)^2 S_{PO}^4 + c_L (p-1) S_O^2 S_{PO}^2,$$

$$V_U = a_U S_O^4 + b_U (p-1)^2 S_{PO}^4 + c_U (p-1) S_O^2 S_{PO}^2,$$

$$a_L = \left[1 - \frac{2}{po(r-1)} - \frac{1}{F_{1-\alpha/2:o-1,po(r-1)}}\right]^2,$$

$$b_L = \left[1 - \frac{2}{po(r-1)} - \frac{1}{F_{1-\alpha/2:(p-1)(o-1),po(r-1)}}\right]^2,$$

$$c_L = \left[1 - \frac{2}{po(r-1)} - \frac{1}{F_{1-\alpha/2:p(o-1),po(r-1)}}\right]^2 \frac{p^2}{p-1} - \frac{a_L}{p-1} - (p-1)b_L,$$

$$a_U = \left[\frac{1}{F_{\alpha/2:o-1,po(r-1)}} - 1 + \frac{2}{po(r-1)}\right]^2,$$

$$b_U = \left[\frac{1}{F_{\alpha/2:(p-1)(o-1),po(r-1)}} - 1 + \frac{2}{po(r-1)}\right]^2, \text{ and}$$

$$c_U = \left[\frac{1}{F_{\alpha/2:p(o-1),po(r-1)}} - 1 + \frac{2}{po(r-1)}\right]^2 \frac{p^2}{p-1} - \frac{a_U}{p-1} - (p-1)b_U.$$

To demonstrate, we will compute 95% confidence intervals for η and ρ_{repeat} using the data in Section 3.5. To begin, we compute

$$
\begin{aligned}
a_L &= \left[1 - \frac{2}{po(r-1)} - \frac{1}{F_{1-\alpha/2:o-1,\,po(r-1)}}\right]^2 \\
&= \left[1 - \frac{2}{120} - \frac{1}{2.6740}\right]^2 \\
&= 0.3713, \\
b_L &= \left[1 - \frac{2}{po(r-1)} - \frac{1}{F_{1-\alpha/2:(p-1)(o-1),\,po(r-1)}}\right]^2, \\
&= \left[1 - \frac{2}{120} - \frac{1}{1.4600}\right]^2, \\
&= 0.08904, \\
c_L &= \left[1 - \frac{2}{po(r-1)} - \frac{1}{F_{1-\alpha/2:p(o-1),\,po(r-1)}}\right]^2 \frac{p^2}{p-1} - \frac{a_L}{p-1} - (p-1)b_L, \\
&= \left[1 - \frac{2}{120} - \frac{1}{1.4536}\right]^2 \frac{400}{19} - \frac{a_L}{19} - 19b_L, \\
&= 0.1254, \\
a_U &= \left[\frac{1}{F_{\alpha/2:o-1,\,po(r-1)}} - 1 + \frac{2}{po(r-1)}\right]^2, \\
&= \left[\frac{1}{0.16476} - 1 + \frac{2}{120}\right]^2, \\
&= 25.87, \\
b_U &= \left[\frac{1}{F_{\alpha/2:(p-1)(o-1),\,po(r-1)}} - 1 + \frac{2}{po(r-1)}\right]^2, \text{ and} \\
&= \left[\frac{1}{0.67907} - 1 + \frac{2}{120}\right]^2, \text{ and} \\
&= 0.2394 \\
c_U &= \left[\frac{1}{F_{\alpha/2:p(o-1),\,po(r-1)}} - 1 + \frac{2}{po(r-1)}\right]^2 \frac{p^2}{p-1} - \frac{a_U}{p-1} - (p-1)b_U \\
&= \left[\frac{1}{0.68348} - 1 + \frac{2}{120}\right]^2 \frac{400}{19} - \frac{a_U}{19} - 19b_U \\
&= -1.064 \\
V_L &= a_L S_O^4 + b_L(p-1)^2 S_{PO}^4 + c_L(p-1)S_O^2 S_{PO}^2, \\
&= a_L(13.59)^2 + b_L(19)^2(2.060)^2 + c_L(19)(13.59)(2.060), \\
&= 271.7,
\end{aligned}
$$

$$\begin{aligned} V_U &= a_U S_O^4 + b_U (p-1)^2 S_{PO}^4 + c_U (p-1) S_O^2 S_{PO}^2, \\ &= a_U (13.59)^2 + b_U (19)^2 (2.060)^2 + c_U (19)(13.59)(2.060), \\ &= 4{,}576. \end{aligned}$$

Thus, the 95% confidence interval for η based on Equation (3.12) is

$$\begin{aligned} L &= \frac{1}{prS_E^2}\left[\left(1-\frac{2}{po(r-1)}\right)\left(S_O^2+(p-1)S_{PO}^2\right)-\sqrt{V_L}\right]-\frac{1}{r} \\ &= \frac{1}{40(0.7333)}\left[\left(1-\frac{2}{120}\right)\left(13.59+(19)2.060\right)-\sqrt{V_L}\right]-\frac{1}{2} \\ &= 0.706 \end{aligned}$$

and

$$\begin{aligned} U &= \frac{1}{prS_E^2}\left[\left(1-\frac{2}{po(r-1)}\right)\left(S_O^2+(p-1)S_{PO}^2\right)+\sqrt{V_U}\right]-\frac{1}{r} \\ &= \frac{1}{40(0.7333)}\left[\left(1-\frac{2}{120}\right)\left(13.59+(19)2.060\right)+\sqrt{V_U}\right]-\frac{1}{2} \\ &= 3.57. \end{aligned}$$

Based on this interval for η and the relation in Equation (3.11), the 95% confidence interval for ρ_{repeat} is

$$\begin{aligned} L &= \frac{1}{3.57+1} \\ &= 0.219 \end{aligned}$$

and

$$\begin{aligned} U &= \frac{1}{0.706+1} \\ &= 0.586. \end{aligned} \tag{3.13}$$

A generalized confidence interval for η is based on the pivotal quantity

$$\mathrm{GPQ}(\eta) = \frac{\frac{(o-1)s_O^2}{W_2} + (p-1)\frac{(p-1)(o-1)s_{PO}^2}{W_3}}{pr\frac{po(r-1)s_E^2}{W_4}} - \frac{1}{r},$$

where s_O^2, s_{PO}^2, and s_E^2 are the realized values of S_O^2, S_{PO}^2, and S_E^2 and W_2, W_3, and W_4 are independent chi-squared random variables with $o-1$, $(p-1)(o-1)$, and $po(r-1)$ degrees of freedom, respectively. The generalized confidence interval for ρ_{repeat} is obtained from the interval for η using Equation (3.11). For the data in this example, we obtain the 95% generalized confidence interval for ρ_{repeat} from $L = 0.210$ to $U = 0.576$ (using 100,000 iterations). This interval is very similar to the one computed in Equation (3.13).

Using results from the generalized confidence interval, we see that repeatability is responsible for at least 21% and at most 57.6% of the measurement error. Conversely, reproducibility is responsible for at least 42.4% and at most 79% of the measurement error. Since reproducibility is likely the greater contributor to measurement error, one should focus on the operators as a starting point to improve the measurement system. For situations in which the measurement system is not capable and repeatability constitutes the majority of the measurement error, it might be prudent to develop a new measurement system instead of investing more time and money on the present system.

3.7 Comparing Two Measurement Systems

It is sometimes of interest to compare two measurement systems. This might be done, for example, after a measurement system has been modified and one wants to determine if the modification improved the system. Let γ_{M1} represent measurement variance of the first system and γ_{M2} represent measurement variance of the second system. Our objective is to compute a confidence interval for the ratio γ_{M1}/γ_{M2}. If all values in the confidence interval for this ratio exceed one, this provides evidence $\gamma_{M1} > \gamma_{M2}$. This suggests that system 2 is better than system 1. Conversely, if all values in the interval are less than one, this provides evidence $\gamma_{M1} < \gamma_{M2}$ and that system 1 is better than system 2. If the confidence interval contains the value one, there is no evidence that the two systems differ.

Burdick, Allen, and Larsen [6] compared two closed-form confidence intervals for γ_{M1}/γ_{M2}. They recommended an interval proposed by Cochran [16] but warned that the confidence coefficient can be less than the stated level if the number of operators is less than six. An alternative MLS interval proposed by Ting, Burdick, and Graybill [63] maintains the stated confidence level in most situations, but it can be much wider than the Cochran interval when there are fewer than six operators. It is also more difficult to compute than the Cochran interval. The formula for the MLS interval is given in the appendix of [6].

The $100(1-\alpha)\%$ confidence interval for γ_{M1}/γ_{M2} recommended by Cochran [16] is

$$L = \frac{\widehat{\gamma}_{M1}}{\widehat{\gamma}_{M2} F_{1-\alpha/2:m_1,m_2}}$$

and

$$U = \frac{\widehat{\gamma}_{M1}}{\widehat{\gamma}_{M2} F_{\alpha/2:m_1,m_2}}, \tag{3.14}$$

where

$$\widehat{\gamma}_{M1} = \frac{S_{O1}^2 + (p_1 - 1)S_{PO1}^2 + p_1(r_1 - 1)S_{E1}^2}{p_1 r_1},$$

$$\widehat{\gamma}_{M2} = \frac{S_{O2}^2 + (p_2 - 1)S_{PO2}^2 + p_2(r_2 - 1)S_{E2}^2}{p_2 r_2},$$

$$m_1 = \frac{p_1^2 r_1^2 \widehat{\gamma}_{M1}^2}{\frac{S_{O1}^4}{o_1 - 1} + \frac{(p_1-1)^2 S_{PO1}^4}{(p_1-1)(o_1-1)} + \frac{p_1^2(r_1-1)^2 S_{E1}^4}{p_1 o_1 (r_1-1)}},$$

$$m_2 = \frac{p_2^2 r_2^2 \widehat{\gamma}_{M2}^2}{\frac{S_{O2}^4}{o_2 - 1} + \frac{(p_2-1)^2 S_{PO2}^4}{(p_2-1)(o_2-1)} + \frac{p_2^2(r_2-1)^2 S_{E2}^4}{p_2 o_2 (r_2-1)}},$$

and the numerical subscripts denote the measurement system (1 or 2).

Table 3.13. *ANOVA for measurement system* 1.

Source of variation	Degrees of freedom	Mean square
Operators (O)	5	9.020
Parts × O	40	1.802
Replicates	54	0.191

Table 3.14. *ANOVA for measurement system* 2.

Source of variation	Degrees of freedom	Mean square
Operators (O)	5	4.144
Parts × O	40	0.235
Replicates	54	0.335

A GPQ that can be used to construct a confidence interval on γ_{M1}/γ_{M2} is

$$\frac{\text{GPQ}(\gamma_{M1})}{\text{GPQ}(\gamma_{M2})}, \tag{3.15}$$

where

$$\text{GPQ}(\gamma_{M1}) = \frac{(o_1-1)s_{O1}^2}{p_1 r_1 W_{21}} + \frac{(p_1-1)^2(o_1-1)s_{PO1}^2}{p_1 r_1 W_{31}} + \frac{p_1 o_1 (r_1-1)^2 s_{E1}^2}{r_1 W_{41}},$$

$$\text{GPQ}(\gamma_{M2}) = \frac{(o_2-1)s_{O2}^2}{p_2 r_2 W_{22}} + \frac{(p_2-1)^2(o_2-1)s_{PO2}^2}{p_2 r_2 W_{32}} + \frac{p_2 o_2 (r_2-1)^2 s_{E2}^2}{r_2 W_{42}},$$

and W_{21}, W_{22}, W_{31}, W_{32}, W_{41}, and W_{42} are independent chi-squared random variables with degrees of freedom o_1-1, o_2-1, $(p_1-1)(o_1-1)$, $(p_2-1)(o_2-1)$, $p_1o_1(r_1-1)$, and $p_2o_2(r_2-1)$, respectively.

Burdick, Allen, and Larsen [6] describe an application concerning the manufacturing of tape drives for computer backup. The tape drive heads are manufactured by an outside supplier and must be tested to ensure they meet performance specifications. A newly developed head tester was installed at the supplier's production facility to test heads before shipment. Tape cartridges are placed in the tester to measure various characteristics of the tape heads. A gauge R&R study was performed to determine whether the head tester was capable of monitoring the manufacturing process. In the study there were $p_1 = 9$ heads (parts), $o_1 = 6$ cartridges (operators), and $r_1 = 2$ replicates. After performing the analysis, an engineer made some changes to the tester and a second gauge R&R study was conducted to determine the effect of these changes. In the second experiment, $p_2 = 9$ parts were used with $o_2 = 6$ and $r_2 = 2$.

Table 3.13 presents the ANOVA for measurement system 1 and Table 3.14 presents the ANOVA for measurement system 2 for the response variable "reverse resolution." (The mean squares for parts are omitted because they are not used in the computations.)

Using Equation (3.14), we first compute

$$\begin{aligned}
\widehat{\gamma}_{M1} &= \frac{S_{O1}^2 + (p_1 - 1)S_{PO1}^2 + p_1(r_1 - 1)S_{E1}^2}{p_1 r_1} \\
&= \frac{9.020 + 8(1.802) + 9(1)(0.191)}{9(2)} \\
&= 1.398, \\
\widehat{\gamma}_{M2} &= \frac{S_{O2}^2 + (p_2 - 1)S_{PO2}^2 + p_2(r_2 - 1)S_{E2}^2}{p_2 r_2} \\
&= \frac{4.144 + 8(0.235) + 9(1)(0.335)}{9(2)} \\
&= 0.5022, \\
m_1 &= \frac{p_1^2 r_1^2 \widehat{\gamma}_{M1}^2}{\frac{S_{O1}^4}{o_1 - 1} + \frac{(p_1-1)^2 S_{PO1}^4}{(p_1-1)(o_1-1)} + \frac{p_1^2 (r_1-1)^2 S_{E1}^4}{p_1 o_1 (r_1 - 1)}} \\
&= \frac{(9)^2(2)^2(1.398)^2}{\frac{(9.020)^2}{5} + \frac{(8)^2(1.802)^2}{40} + \frac{(9)^2(1)^2(0.191)^2}{54}} \\
&= 29 \text{ (truncated)},
\end{aligned}$$

and

$$\begin{aligned}
m_2 &= \frac{p_2^2 r_2^2 \widehat{\gamma}_{M2}^2}{\frac{S_{O2}^4}{o_2 - 1} + \frac{(p_2-1)^2 S_{PO2}^4}{(p_2-1)(o_2-1)} + \frac{p_2^2 (r_2-1)^2 S_{E2}^4}{p_2 o_2 (r_2 - 1)}} \\
&= \frac{(9)^2(2)^2(0.5022)^2}{\frac{(4.144)^2}{5} + \frac{(8)^2(0.235)^2}{40} + \frac{(9)^2(1)^2(0.335)^2}{54}} \\
&= 22 \text{ (truncated)}.
\end{aligned}$$

Using these values, the 95% Cochran confidence interval for γ_{M1}/γ_{M2} is

$$\begin{aligned}
L &= \frac{\widehat{\gamma}_{M1}}{\widehat{\gamma}_{M2} F_{0.975:29,22}} \\
&= \frac{1.398}{0.5022(2.280)} \\
&= 1.2
\end{aligned}$$

and

$$\begin{aligned}
U &= \frac{\widehat{\gamma}_{M1}}{\widehat{\gamma}_{M2} F_{0.025:29,22}} \\
&= \frac{1.398}{0.5022(0.4585)} \\
&= 6.1. \qquad (3.16)
\end{aligned}$$

The 95% generalized confidence interval for γ_{M1}/γ_{M2} based on 100,000 of the GPQ values shown in Equation (3.15) is from $L = 0.87$ to $U = 7.8$. Here the interval in Equation (3.16) is much shorter than the generalized confidence interval. This is generally the case, and since there are six operators in the experiment, we expect the Cochran interval to maintain the stated confidence level. Since all values in the Cochran interval are greater than one, it seems reasonable to conclude the modification has improved the measurement system.

3.8 Intervals for Other Parameters

As discussed in the preface, one objective of this book is to provide a comprehensive reference of confidence intervals for any mixed or random model. The parameters of interest in a gauge R&R study were presented in previous sections of this chapter. In this section, we provide results for parameters that are sometimes of interest in other applications. The discussion is brief, and readers interested in more detail should consult Burdick and Graybill [10, pp. 118–126]. (Note that Burdick and Graybill use F_α to represent an F-percentile with area α to the *right*. In this book, F_α represents an F-percentile with area α to the *left*.) Table 3.15 reports constants not defined in Table 3.7 that are used in the formulas that follow.

Table 3.15. *Additional constants used for confidence intervals in Chapter 3. Values are for $\alpha = 0.05$, $p = 20$, $o = 6$, and $r = 2$.*

Constant	Definition	Value
F_5	$F_{1-\alpha/2:o-1,(p-1)(o-1)}$	2.703
F_6	$F_{\alpha/2:o-1,(p-1)(o-1)}$	0.1644
F_7	$F_{1-\alpha/2:(p-1)(o-1),po(r-1)}$	1.460
F_8	$F_{\alpha/2:(p-1)(o-1),po(r-1)}$	0.6791
F_9	$F_{1-\alpha/2:p-1,po(r-1)}$	1.845
F_{10}	$F_{\alpha/2:p-1,po(r-1)}$	0.4539
F_{11}	$F_{1-\alpha/2:o-1,po(r-1)}$	2.674
F_{12}	$F_{\alpha/2:o-1,po(r-1)}$	0.1648
G_{23}	$\dfrac{(F_5-1)^2 - G_2^2F_5^2 - H_3^2}{F_5}$	0.01843
H_{23}	$\dfrac{(1-F_6)^2 - H_2^2F_6^2 - G_3^2}{F_6}$	−0.2173
G_{34}	$\dfrac{(F_7-1)^2 - G_3^2F_7^2 - H_4^2}{F_7}$	−0.000346
H_{34}	$\dfrac{(1-F_8)^2 - H_3^2F_8^2 - G_4^2}{F_8}$	−0.001597

3.8.1 Intervals for σ_O^2, σ_{PO}^2, and σ_E^2

Confidence intervals for $\sigma_O^2 = (\theta_O - \theta_{PO})/(pr)$ and $\sigma_{PO}^2 = (\theta_{PO} - \theta_E)/r$ are computed using the method described in Section 3.3.2 for σ_P^2 with appropriate substitutions. In particular, the bounds of an approximate $100(1-\alpha)\%$ confidence interval for σ_O^2 are

$$L = \widehat{\sigma}_O^2 - \frac{\sqrt{V_{LO}}}{pr}$$

and

$$U = \widehat{\sigma}_O^2 + \frac{\sqrt{V_{UO}}}{pr}, \tag{3.17}$$

where

$$\widehat{\sigma}_O^2 = \frac{S_O^2 - S_{PO}^2}{pr},$$

$$V_{LO} = G_2^2 S_O^4 + H_3^2 S_{PO}^4 + G_{23} S_O^2 S_{PO}^2,$$

and

$$V_{UO} = H_2^2 S_O^4 + G_3^2 S_{PO}^4 + H_{23} S_O^2 S_{PO}^2. \tag{3.18}$$

The bounds of an approximate $100(1-\alpha)\%$ confidence interval for σ_{PO}^2 are

$$L = \widehat{\sigma}_{PO}^2 - \frac{\sqrt{V_{LPO}}}{r}$$

and

$$U = \widehat{\sigma}_{PO}^2 + \frac{\sqrt{V_{UPO}}}{r}, \tag{3.19}$$

where

$$\widehat{\sigma}_{PO}^2 = \frac{S_{PO}^2 - S_E^2}{r},$$

$$V_{LPO} = G_3^2 S_{PO}^4 + H_4^2 S_E^4 + G_{34} S_{PO}^2 S_E^2,$$

$$V_{UPO} = H_3^2 S_{PO}^4 + G_4^2 S_E^4 + H_{34} S_{PO}^2 S_E^2,$$

G_2, G_3, G_4, H_2, H_3, and H_4 are defined in Table 3.7, and G_{23}, H_{23}, G_{34}, and H_{34} are defined in Table 3.15. Negative bounds are increased to zero for both Equation (3.17) and Equation (3.19).

An exact test of size $\alpha/2$ for the null hypothesis $H_0 : \sigma_{PO}^2 = 0$ versus $H_a : \sigma_{PO}^2 > 0$ is to reject H_0 if $S_{PO}^2/S_E^2 > F_7$. If $S_{PO}^2/S_E^2 < F_7$, then L in Equation (3.19) is negative. In such a situation, an investigator may decide to fit a two-factor model with no interaction. This model is discussed in Chapter 5.

The $100(1-\alpha)\%$ confidence interval for σ_E^2 using Result 5 in Table 3.4 is

$$L = (1 - G_4) S_E^2$$

and

$$U = (1 + H_4) S_E^2, \tag{3.20}$$

where G_4 and H_4 are defined in Table 3.7.

3.8.2 Interval for γ_Y

The total variation in the response variable is $\gamma_Y = \sigma_P^2 + \sigma_O^2 + \sigma_{PO}^2 + \sigma_E^2$. This is written in terms of expected mean squares as $\gamma_Y = [p\theta_P + o\theta_O + (po - p - o)\theta_{PO} + po(r-1)\theta_E]/(por)$. Since this a sum of expected mean squares, we apply the MLS method proposed by Graybill and Wang [25]. The resulting bounds of an approximate $100(1-\alpha)\%$ confidence interval for γ_Y are

$$L = \widehat{\gamma}_Y - \frac{\sqrt{V_{LT}}}{por}$$

and

$$U = \widehat{\gamma}_Y + \frac{\sqrt{V_{UT}}}{por},$$

where

$$\widehat{\gamma}_Y = \frac{pS_P^2 + oS_O^2 + (po - p - o)S_{PO}^2 + po(r-1)S_E^2}{por},$$

$$V_{LT} = G_1^2 p^2 S_P^4 + G_2^2 o^2 S_O^4 + G_3^2 (po - p - o)^2 S_{PO}^4 + G_4^2 (po)^2 (r-1)^2 S_E^4,$$

$$V_{UT} = H_1^2 p^2 S_P^4 + H_2^2 o^2 S_O^4 + H_3^2 (po - p - o)^2 S_{PO}^4 + H_4^2 (po)^2 (r-1)^2 S_E^4,$$

and G_1, G_2, G_3, G_4, H_1, H_2, H_3, and H_4 are defined in Table 3.7.

3.8.3 Intervals for σ_P^2/σ_E^2, σ_O^2/σ_E^2, and σ_{PO}^2/σ_E^2

It is often of interest to compare the magnitude of each variance component to the error component, σ_E^2. These ratios expressed in terms of the expected mean squares are $\sigma_P^2/\sigma_E^2 = (\theta_P - \theta_{PO})/(or\theta_E)$, $\sigma_O^2/\sigma_E^2 = (\theta_O - \theta_{PO})/(pr\theta_E)$, and $\sigma_{PO}^2/\sigma_E^2 = (\theta_{PO} - \theta_E)/(r\theta_E)$. The confidence intervals for σ_P^2/σ_E^2 and σ_O^2/σ_E^2 are based on a method proposed by Wang and Graybill [70]. In particular, the $100(1-\alpha)\%$ confidence interval for σ_P^2/σ_E^2 is

$$L = \frac{S_{PO}^2}{orS_E^2F_9}\left[\frac{S_P^2}{S_{PO}^2} - \frac{1}{1-G_1} + \frac{S_{PO}^2F_1(1 - F_1(1-G_1))}{S_P^2(1-G_1)}\right]$$

and

$$U = \frac{S_{PO}^2}{orS_E^2F_{10}}\left[\frac{S_P^2}{S_{PO}^2} - \frac{1}{1+H_1} + \frac{S_{PO}^2F_2(1 - F_2(1+H_1))}{S_P^2(1+H_1)}\right].$$

The $100(1-\alpha)\%$ confidence interval for σ_O^2/σ_E^2 is

$$L = \frac{S_{PO}^2}{prS_E^2F_{11}}\left[\frac{S_O^2}{S_{PO}^2} - \frac{1}{1-G_2} + \frac{S_{PO}^2F_5(1 - F_5(1-G_2))}{S_O^2(1-G_2)}\right]$$

and

$$U = \frac{S_{PO}^2}{prS_E^2F_{12}}\left[\frac{S_O^2}{S_{PO}^2} - \frac{1}{1+H_2} + \frac{S_{PO}^2F_6(1 - F_6(1+H_2))}{S_O^2(1+H_2)}\right],$$

where G_1, G_2, H_1, H_2, F_1, and F_2 are defined in Table 3.7 and F_5, F_6, F_9, F_{10}, F_{11}, and F_{12} are defined in Table 3.15.

An exact confidence interval for σ^2_{PO}/σ^2_E is formed using Results 1, 4, and 5 of Table 3.4. From these results, $(S^2_{PO}\theta_E)/(S^2_E\theta_{PO})$ has an exact F-distribution with $(p-1)(o-1)$ and $po(r-1)$ degrees of freedom. This provides the exact $100(1-\alpha)\%$ confidence interval on σ^2_{PO}/σ^2_E,

$$L = \frac{1}{r}\left[\frac{S^2_{PO}}{S^2_E F_7} - 1\right]$$

and

$$U = \frac{1}{r}\left[\frac{S^2_{PO}}{S^2_E F_8} - 1\right], \tag{3.21}$$

where F_7 and F_8 are defined in Table 3.15.

3.8.4 Intervals for σ^2_O/γ_Y, σ^2_{PO}/γ_Y, and σ^2_E/γ_Y

Ratios of individual variance components to the total variability are of interest in genetics and other fields of application. Leiva and Graybill [42] derived the $100(1-\alpha)\%$ interval for σ^2_O/γ_Y,

$$L_O = \frac{L^*_O}{L^*_O + 1}$$

and

$$U_O = \frac{U^*_O}{U^*_O + 1}, \tag{3.22}$$

where

$$L^*_O = \max\left[0, \frac{o(1-G_2)(S^2_O - F_5 S^2_{PO})}{po(r-1)S^2_E + p(1-G_2)S^2_P/F_4 + p(o-1)S^2_{PO}}\right]$$

and

$$U^*_O = \max\left[0, \frac{o(1+H_2)(S^2_O - F_6 S^2_{PO})}{po(r-1)S^2_E + p(1+H_2)S^2_P/F_3 + p(o-1)S^2_{PO}}\right].$$

Leiva and Graybill [42] also proposed the $100(1-\alpha)\%$ interval for σ^2_{PO}/γ_Y,

$$L_{PO} = \frac{poL^*_{PO}}{L^*_{PO}(po-p-o)+po}$$

and

$$U_{PO} = \frac{poU^*_{PO}}{U^*_{PO}(po-p-o)+po}, \tag{3.23}$$

where

$$L^*_{PO} = \max\left[0, \frac{poF_2F_6(S^2_{PO} - F_7 S^2_E)}{pF_6S^2_P + oF_2S^2_O + (por-p-o)F_2F_6F_7S^2_E}\right],$$

$$U^*_{PO} = \max\left[0, \frac{poF_1F_5(S^2_{PO} - F_8 S^2_E)}{pF_5S^2_P + oF_1S^2_O + (por-p-o)F_1F_5F_8S^2_E}\right],$$

Table 3.16. *GPQs for other parameters in model* (3.1).

Parameter	GPQ
σ_O^2	$\max\left[0, \dfrac{(o-1)s_O^2}{prW_2} - \dfrac{(p-1)(o-1)s_{PO}^2}{prW_3}\right]$
σ_{PO}^2	$\max\left[0, \dfrac{(p-1)(o-1)s_{PO}^2}{rW_3} - \dfrac{po(r-1)s_E^2}{rW_4}\right]$
γ_Y	$\dfrac{(p-1)s_P^2}{orW_1} + \dfrac{(o-1)s_O^2}{prW_2} + \dfrac{(po-p-o)(p-1)(o-1)s_{PO}^2}{porW_3} + \dfrac{po(r-1)^2 s_E^2}{rW_4}$
σ_P^2/σ_E^2	$\max\left[0, \dfrac{(p-1)s_P^2/W_1 - (p-1)(o-1)s_{PO}^2/W_3}{po^2r(r-1)s_E^2/W_4}\right]$
σ_O^2/σ_E^2	$\max\left[0, \dfrac{(o-1)s_O^2/W_2 - (p-1)(o-1)s_{PO}^2/W_3}{p^2or(r-1)s_E^2/W_4}\right]$
σ_O^2/γ_Y	$\dfrac{\text{GPQ}(\sigma_O^2)}{\text{GPQ}(\gamma_Y)}$
σ_{PO}^2/γ_Y	$\dfrac{\text{GPQ}(\sigma_{PO}^2)}{\text{GPQ}(\gamma_Y)}$
σ_E^2/γ_Y	$\dfrac{po(r-1)s_E^2/W_4}{\text{GPQ}(\gamma_Y)}$

and G_2, H_2, F_1, F_2, F_3, and F_4 are defined in Table 3.7, and F_5, F_6, F_7, and F_8 are defined in Table 3.15.

The MLS confidence interval for σ_E^2/γ_Y is more complex than Equations (3.22) and (3.23). Readers interested in this formulation are referred to [10, p. 125].

GPQs for the parameters in this section are shown in Table 3.16, where W_1, W_2, W_3, and W_4 are jointly independent chi-squared random variables with degrees of freedom $p-1$, $o-1$, $(p-1)(o-1)$, and $po(r-1)$, respectively. The terms s_P^2, s_O^2, s_{PO}^2, and s_E^2 are the realized values of the mean squares for a particular data set. Since the intervals in Equations (3.20) and (3.21) are exact, no GPQs are reported for σ_E^2 and σ_{PO}^2/σ_E^2.

Table 3.17 reports the computed intervals for the example data presented in Section 3.5. The generalized confidence intervals are based on 100,000 simulated GPQ values.

3.9 Summary

In this chapter we presented results for the most common type of gauge R&R study. In many applications, there is no evidence that an interaction effect between parts and operators exists. In such cases, a more appropriate model is the two-factor crossed model with no

Table 3.17. *Computed intervals for other parameters in example. (See Section 1.8 for a description of computer programs to perform these computations.)*

Parameter	MLS	GCI
σ_O^2	$L = 0.0792$	$L = 0.0798$
	$U = 1.99$	$U = 2.01$
σ_{PO}^2	$L = 0.398$	$L = 0.395$
	$U = 1.04$	$U = 1.04$
σ_E^2	$L = 0.578$	Exact
	$U = 0.961$	
γ_Y	$L = 30.0$	$L = 30.2$
	$U = 107$	$U = 107$
σ_P^2/σ_E^2	$L = 36.2$	$L = 36.2$
	$U = 148$	$U = 148$
σ_O^2/σ_E^2	$L = 0.104$	$L = 0.105$
	$U = 2.74$	$U = 2.76$
σ_{PO}^2/σ_E^2	$L = 0.462$	Exact
	$U = 1.57$	
σ_O^2/γ_Y	$L = 0.00120$	$L = 0.00130$
	$U = 0.0399$	$U = 0.0401$
σ_{PO}^2/γ_Y	$L = 0.00438$	$L = 0.00531$
	$U = 0.0284$	$U = 0.0259$
σ_E^2/γ_Y		$L = 0.00662$
		$U = 0.0259$

interaction. This model is discussed in Chapter 5. Modifications to the results in this chapter when operators are fixed are given in Chapter 6. Chapter 7 considers the two-factor model when replications are not equal for all treatment combinations. In the next chapter we present guidelines for designing a gauge R&R study.

Chapter 4

Design of Gauge R&R Experiments

4.1 Introduction

In Chapters 2 and 3 we introduced two basic experimental designs for conducting gauge R&R studies. These designs are the single-factor design and the two-factor crossed (factorial) design. The objective for this chapter is to review some of the basic principles of designing experiments and to discuss how those principles apply to gauge R&R studies.

Montgomery [51] lists three basic principles of designing experiments: randomization, replication, and blocking. While these are very general principles that have application to all experiments, they can be particularly important in gauge R&R studies as well as other more general types of measurement systems capability studies.

Replication is a repetition of the basic experiment. In the gauge R&R experiment described in Chapter 3 (Table 3.1), replication consists of an operator making a measurement on a part. Two measurements are made by each operator on a part, and replication assumes that these measurements are made independently of each other with unique setup of any tool or fixture that is required. Replication has two important properties. First, it allows the experimenter to obtain an estimate of the experimental error (σ_E^2 in Chapter 3). This estimate of error is the repeatability of the measuring device. It also defines the scale that we use for determining whether observed differences in the data are really statistically different. Second, replication improves the precision with which model parameters are estimated. For example, suppose the repeatability variance component σ_E^2 is of interest. Replication permits the experimenter to obtain a more precise estimate of this effect because the width of the confidence interval depends on the number of replicates.

There is an important distinction between replication and repeated measurements. Suppose that a part is measured three consecutive times by the operator without changing the setup of the measuring device or any fixturing of the part that may be necessary. The measurements so obtained are not replicates. Rather, they are a form of repeated measurements. Furthermore, since the operator knows that he or she is measuring the same part, there may be a tendency for the operator to report all measurements as the same value. This can result in significant underestimation of gauge repeatability.

Randomization is the cornerstone underlying the use of statistical methods in experimental design. By randomization we mean that both the allocation of the experimental material and the order in which the individual runs or trials of the experiment are to be performed are randomly determined. In a gauge R&R experiment, this means that each operator should measure all p parts in random order. Statistical methods require that the observations are independently distributed random variables. Randomization plays an important role in making this assumption valid. By properly randomizing the experiment, we also assist in averaging out the effects of extraneous factors that may be present. For example, suppose that in the experiment shown in Table 3.1 there is a warm-up effect on the instrument used to measure thermal impedance such that the measurements drift higher the longer the instrument is on. By making the observations in random order, we minimize any potential bias that could result from the drift in the measurement instrument.

Computer software programs are often used to assist experimenters in selecting and constructing experimental designs. Typically these programs present the runs in the experimental design in random order. This random order is usually created from an internal random number generator. Even with such a computer program, it is still often necessary to assign experimental material (such as parts) or the gauges used in the experiment. It is important to ensure that all such assignments are random.

Sometimes experimenters encounter situations in which complete randomization of some aspect of the experiment is difficult. For example, part fixturing or measurement tool setup may be a hard-to-change variable, making complete randomization of this factor very difficult. Statistical design methods based on split-plots could be used for dealing with restrictions on randomization. However, very little use of split-plot designs has been reported in the literature on measurement system capability studies. For basic information about the split-plot design, see Montgomery [51, Chapter 13].

Blocking is a design technique used to improve the precision with which comparisons among the factors of interest are made. Often blocking is used to reduce or eliminate variability transmitted from nuisance factors. These are factors that may influence the experimental response but in which we are not directly interested. For example, a measurement systems capability study on an analytical chemical procedure or assay may require two batches of raw material to make all the required runs. However, there could be differences between the batches due to supplier-to-supplier variability, and if we are not specifically interested in this effect, we would consider the batches of raw material as a nuisance factor. Generally, a block is a set of relatively homogeneous experimental conditions. In the chemical assay example, each batch of raw material would form a block, because the variability within a batch would be expected to be smaller than the variability between batches. Typically, as in this example, each level of the nuisance factor becomes a block. Then the experimenter divides the observations from the statistical design into groups that are run in each block.

4.2 Guidelines for Designing Experiments

Montgomery [51] identifies a seven-step process for successful design of an experiment. The steps are summarized here in Table 4.1. Steps 1, 2, and 3 are the preexperimental planning activities. Steps 2 and 3 are often performed simultaneously, or occasionally the order may be reversed.

Table 4.1. *Guidelines for designing an experiment.*

1.	Recognition and statement of the problem.
2.	Choice of factors, levels, and ranges.
3.	Selection of the response variable.
4.	Choice of experimental design.
5.	Performing the experiment.
6.	Statistical analysis of the data.
7.	Conclusions and recommendations.

1. **Recognition and statement of the problem.** This step seems rather obvious, but in many problems it is often not easy to realize that a problem requiring experimentation exists. It is also not simple to develop a clear and generally accepted statement of the objectives of the specific experiment that is going to be performed. It is necessary to develop all ideas about the objectives of the experiment. Usually, it is important to solicit input from all concerned parties: engineering, quality assurance, manufacturing, marketing, management, the customer, and operating personnel (who usually have much insight and who are too often ignored). For this reason a team approach to designing experiments is recommended. This can be very important in measurement system capability experiments, where input from all sources knowledgeable about the gauge or measurement system and its use should be solicited. Whether the objective is to obtain confidence interval estimates of variance components or to evaluate the probabilities of part misclassification, the objective should be clear from the beginning.

2. **Choice of factors, levels, and ranges.** (As noted previously, steps 2 and 3 are often done simultaneously, or in the reverse order.) When considering the factors that may influence the performance of a process or system, the experimenter usually discovers that these factors can be classified as either potential design factors or nuisance factors. The potential design factors are those factors that the experimenter may wish to vary in the experiment. Often we find that there are many potential design factors, and some further classification of them is helpful. Some useful classifications are design factors, held-constant factors, and allowed-to-vary factors. The design factors are the factors actually selected for study in the experiment. Held-constant factors are variables that may exert some effect on the response, but for purposes of the present experiment these factors are not of interest, so they will be held at a specific level. For example, in a gauge R&R study the performance of the gauge may be affected by ambient temperature but the experimenter may decide to perform all experimental runs at one controlled temperature level. Thus this factor has been held constant. Nuisance factors, on the other hand, may have large effects that must be accounted for, yet we may not be interested in them in the context of the present experiment. Nuisance factors are often classified as controllable, uncontrollable, or noise factors. A controllable nuisance factor is one whose levels may be set by the experimenter. For example, the experimenter can select different batches of raw material or different days of the week when conducting the experiment. Blocking is often useful in dealing with

controllable nuisance factors. If a nuisance factor is uncontrollable in the experiment, but it can be measured, an analysis procedure called the analysis of covariance often can be used to compensate for its effect. For example, the relative humidity in the process environment may affect gauge performance, and if the humidity cannot be controlled, it probably can be measured and treated as a covariate. When a factor that varies naturally and uncontrollably in the process can be controlled for purposes of an experiment, we often call it a noise factor. In such situations, our objective is usually to find the settings of the controllable design factors that minimize the variability transmitted from the noise factors. This is sometimes called a process robustness study or a robust design problem. Blocking, analysis of covariance, and process robustness studies are discussed in detail in Montgomery [51].

3. **Selection of the response variable.** In selecting the response variable, the experimenter should be certain that this variable provides useful information about the process under study. In most gauge R&R studies, the response variable is relatively easy to define. However, multiple responses are not unusual and should always be considered. Calibration or setup procedures for the gauge are also important, along with standard operating procedures. Operator training and qualification may be a concern. It is important to ensure that these factors are fully considered in designing the experiment.

4. **Choice of experimental design.** If the preexperimental planning activities above are done correctly, this step is usually relatively straightforward. Choice of design involves the consideration of sample size (number of replicates), the selection of a suitable run order for the experimental trials, and the determination of whether blocking or other randomization restrictions are involved. Several interactive statistical software packages support this phase of experimental work. The experimenter can enter information about the number of factors, levels, and ranges, and these programs will either present a selection of designs for consideration or recommend a particular design. (We prefer to see several alternatives instead of relying on a computer recommendation in most cases.) These programs will usually also provide a worksheet (with the order of the runs randomized) for use in collecting the responses.

   In selecting the design, it is important to keep the experimental objectives in mind. In many experiments, we know at the outset that some of the factor levels will result in different values for the response. Consequently, we are interested in identifying which factors cause this difference and in estimating the magnitude of the response change. This is usually the case in gauge R&R experiments, where we expect the gauge to be able to discriminate between different parts. In other situations, we may be more interested in verifying uniformity. For example, we may wish to determine if all operators are equally capable in using a particular type of gauge.

5. **Performing the experiment.** When running the experiment, it is vital to monitor the activity carefully to ensure that everything is being done according to plan. Errors in experimental procedure at this stage will usually destroy experimental validity. Up-front planning is crucial to success. It is easy to underestimate the logistical and planning aspects of running a designed experiment in a complex manufacturing or research and development environment. Before conducting the complete experiment,

a few trial runs or pilot runs often are helpful. These runs provide information about consistency of experimental material, a check on the measurement system, a rough idea of experimental error, and a chance to practice the overall experimental technique. This also provides an opportunity to revisit the decisions made in steps 1–4, if necessary.

6. **Statistical analysis of the data.** Statistical methods should be used to analyze the data so that results and conclusions are objective rather than judgmental. If the experiment has been designed correctly and if it has been performed according to the design, the statistical methods required are not elaborate. This book summarizes the statistical methods necessary for the analysis of gauge R&R experiments. During this phase of the experiment it is important to check the adequacy of the model that has been fit to the data and determine if any of the underlying assumptions have been seriously violated. This topic will be discussed in more detail later in this chapter.

7. **Conclusions and recommendations.** Once the data have been analyzed, the experimenter must draw practical conclusions about the results and recommend a course of action. Sometimes graphical methods are useful in this stage, particularly in presenting the results to others. Follow-up runs and confirmation testing may also be necessary to validate the conclusions from the experiment.

4.3 Factor Structures in Gauge R&R Experiments

Two common factor structures are widely encountered in gauge R&R studies. The first of these, the crossed structure, was introduced in Chapter 3 and illustrated in Table 3.1. Another name for this type of experiment is a factorial design. When two factors are arranged in a crossed or factorial structure, every level of one factor is run in combination with every level of the other factor. Note from Table 3.1 that every part is measured by every operator. The factorial design allows the interaction term between the two factors to be estimated (e.g., the term $(PO)_{ij}$ in Equation (3.1)). Factorial or crossed arrangements can be extended to more than two factors. For example, if there are three factors, a factorial experiment would require that all possible combinations of these factors across all possible levels be run. If there are a levels of factor A, b levels of factor B, and c levels of factor C, the factorial experiment contains abc combinations of factor levels. The model for the experiment would consist of an overall mean, the main effects of all three factors, the three two-factor interactions AB, AC, and BC, a three-factor or ABC interaction term, and the error term. A three-factor crossed design is presented in Section 8.4.

The other type of factor structure that occurs in gauge R&R experiments is the nested arrangement. When two factors are nested, it means that the levels of one factor are similar but not identical to each other at different levels of the other factor. To illustrate, consider the gauge R&R experiment in Table 3.1. This is a factorial experiment because every part is measured by every operator. That is, when an operator measures part 1, it is physically the same part 1 measured by all the operators. Suppose that in conducting this experiment, each operator measures a different set of 20 parts. Now all these parts are randomly chosen from the same (or equivalent) process, so they are similar to each other but are not identical. Consequently, if the experiment had been conducted this way, the factor "parts" is nested within the factor "operators." It might have been necessary to conduct the experiment this way because the operators are in different locations or because the number of measurements that can be taken on each part are limited.

When two factors have a nested arrangement, it means that there is no interaction term between the two factors. For example, if the parts were nested in Table 3.1, the model for the experiment would have been written as

$$Y_{ijk} = \mu_Y + O_i + P(O)_{j(i)} + E_{ijk},$$
$$i = 1, \ldots, o, \quad j = 1, \ldots, p, \quad k = 1, \ldots, r,$$

where the term $P(O)_{j(i)}$ is the "parts nested within operators" effect. The general approach to the statistical analysis of nested designs is discussed in [51]. In Section 8.3 we discuss the nested experimental design in the context of gauge R&R experiments.

It is also possible to have experiments that contain a combination of nested and factorial (or crossed) factors. For example, suppose that there are three factors A, B, and C with A and B crossed and C nested within A. The model would contain terms for the main effects of A and B, the AB interaction, the C(A) nested effect, and the BC(A) interaction. There cannot be either an AC or an ABC interaction, because the levels of factor C are nested within the levels of factor A. The statistical analysis of nested-factorial designs is discussed and illustrated in [51]. An example of a nested-factorial gauge R&R experiment is provided in Section 8.5.

Choosing the design for a gauge R&R experiment involves choosing the number of parts, the number of operators, and the number of replicates. Several relatively standard templates for gauge R&R experiments have appeared in the literature. The most popular uses 10 parts, 3 operators, and 2 or 3 replicates. As demonstrated by Burdick and Larsen [11] and Vardeman and VanValkenburg [67], these designs can lead to relatively wide confidence intervals on the gauge R&R parameters. It is very difficult to obtain a reasonably short confidence interval for any function of σ_O^2 with only three operators. Increasing the number of parts or the number of replicates has only minimal impact on the length of confidence intervals for functions of σ_O^2. Therefore, we recommend at least six operators in any gauge R&R study with random operators. Between 10 and 20 parts is usually a reasonable choice, and 2 or 3 replicates will produce reasonable confidence intervals for most other variance components. However, short confidence intervals for the misclassification rates may require as many as 100 parts in some situations. In manual gauging applications, this can be problematic, but in automated measurement and inspection systems, this may present no significant problems.

4.4 Model Adequacy Checking

The decomposition of the total variability in the observations through an ANOVA is purely algebraic. However, the use of the ANOVA to construct confidence intervals requires that certain assumptions on the observations be satisfied. In general, ANOVA procedures are relatively robust to these underlying assumptions. In experiments where some or all of the factors are random (as is frequently the case in gauge R&R experiments), the assumptions are more critical. The specific assumptions of concern are that the model chosen by the experimenter for the analysis is correct and that the repeatability error terms (e.g., E_{ijk}) are normally and independently distributed with mean zero and constant but unknown variance σ_E^2. We denote this assumption as NID(0,σ_E^2). Additionally, when random factors are present, there are normality, independence, and constant variance assumptions for these model terms as well.

In practice these assumptions usually will not hold exactly. Consequently, it is usually unwise to rely on the ANOVA results without carefully checking the validity of these assumptions. Violations of the basic assumptions and model adequacy can be easily investigated by the examination of residuals.

A residual is defined as the difference between an observed value of the response variable Y and the fitted value of this observation based on the underlying statistical model. For example, consider the two-factor crossed gauge R&R experiment in Table 3.1. The residuals would be found from

$$e_{ijk} = Y_{ijk} - \widehat{Y}_{ijk},$$

where e_{ijk} is the residual and $\widehat{Y}_{ijk}$ is the fitted value of the observation Y_{ijk}. The fitted values are usually obtained from a least-squares fit of the assumed statistical model. For the two-factor crossed experiment with interaction, it can be shown (see [51]) that the least-squares estimate of the observation Y_{ijk} is the average of the r observations in the ijth cell, $\overline{Y}_{ij*}$. To illustrate the calculations, consider Operator 1 and Part 1 in Table 3.1. The observations in this cell are 45 and 44, so the cell average is 44.5. Therefore, the residuals are 0.5 and −0.5. The residuals for the other cells are calculated similarly.

The normality assumption concerning E_{ijk} can be checked by plotting a histogram of the residuals. If the NID(0,σ_E^2) assumption on the errors is satisfied, this plot should look like a sample from a normal distribution centered at zero. Another extremely useful procedure is to construct a normal probability plot of the residuals. If the underlying error distribution is normal, the residuals on this plot will fall along a straight line. In visualizing the straight line, remember to place more emphasis on the central values of the plot than on the extremes. Unfortunately, with small samples considerable fluctuation in these plots can occur, so the appearance of a moderate departure from normality does not necessarily imply a serious violation of the assumptions. In experiments with random effects, moderate to severe departures from normality are potentially serious and require further analysis. Gross outliers also will usually be evident on a normal probability plot.

Plotting the residuals in time order of data collection is helpful in detecting correlation among the residuals. A tendency to have runs of positive and negative residuals indicates positive correlation. This would imply that the independence assumption on the errors has been violated. This is a potentially serious problem and one that is difficult to correct, so it is important to prevent the problem if possible when the data are collected. Proper randomization of the experiment is an important step in obtaining independence.

Sometimes the skill of the operators may change as the experiment progresses, or the measurement system may "drift" or become more erratic. This can result in a change in the error variance over time. This condition often results in a plot of residuals versus time that exhibits more spread at one end than at the other, leading to what is usually called a funnel-shaped pattern. Nonconstant variance is a potentially serious problem, particularly in the random effects model. Plotting the residuals versus the operators is also a useful diagnostic. If one of the operators is more erratic than the others, this will usually show up on this plot.

Plotting the residuals versus the parts may also prove informative, particularly if there is a strong interaction between parts and operators. The inability of operators to measure all parts consistently, which shows up in the interaction, may extend to the variability with which the parts are measured, and this may show up on this plot.

Generally, if the model is correct and if the assumptions are satisfied, the residuals should be structureless. Specifically, they should be unrelated to any other variable, including the predicted response. A simple check is to plot the residuals versus the fitted values. This plot should not reveal any obvious pattern. Problems with nonconstant variance also are indicated by a funnel-shaped pattern on this plot. This problem occurs in many gauge R&R experiments because the variance of the observations increases as the magnitude of the measured response increases. This would be the case if the inherent measurement error in the instrument or gauge or the background noise is a constant percentage of the size of the measured response. (This commonly happens with many measuring instruments. The error experienced in the measurement varies directly with the magnitude of the scale reading.) If this were the case, the residuals would get larger as the observed response gets larger, and the plot of residuals versus the fitted values would give the appearance of an outward-opening funnel or megaphone. Nonconstant variance also arises in cases where the observations have a skewed distribution because in skewed distributions the variance tends to be a function of the mean. Remedial actions for this situation can be found in [51].

Figure 4.1 presents four residual plots for the gauge R&R experiment in Table 3.1. These plots include a histogram and normal probability plot of the residuals, a plot of residuals versus the fitted values, and a plot of the residuals in time order. (We do not know the real order of data collection because it was not reported by the original experimenters, so this plot was constructed assuming a particular order of experimentation.) The plot of residuals versus operators is shown in Figure 4.2, and the plot of residuals versus parts is in Figure 4.3. None of these plots reveals any unusual patterns or diagnostic information.

In addition to the standard residual plots that we have discussed, it is also useful to analyze the data from gauge experiments using control charts. These charts provide another way to examine the stability and consistency of the data over the time that the experiment was conducted. Since the typical gauge R&R experiment is a replicated experiment, $\overline{X}$ and R control charts are the proper type of control chart to employ. (See [52] for a description of $\overline{X}$ and R control charts.)

Figure 4.4 shows the $\overline{X}$ and R control charts for the gauge R&R experiment from Table 3.1. In constructing the charts, we have identified each part as a sample, with two observations on each part for each one of the six operators. This produces the 120 samples (each of size two) shown on the horizontal axis of the control chart. Samples 1–20 are from Operator 1, samples 21–40 are from Operator 2, and so on. The center line of the R chart is the average range of the 120 samples. Consequently, it reflects the within-sample variability, which is attributable only to gauge repeatability. The limits on the $\overline{X}$ control chart are determined using the average range. Notice that almost all of the points on the $\overline{X}$ control chart are outside the control limits (what is usually called an out-of-control condition). This indicates that the gauge is capable of discriminating between different parts, and this is actually a desirable finding. If most of the samples on this control chart were in control, it would indicate that it is difficult for the gauge to clearly identify different parts. The samples on the R chart exhibit statistical control, as all samples are inside the control limits. This is also desirable, as it indicates that the operators exhibit consistency in their use of

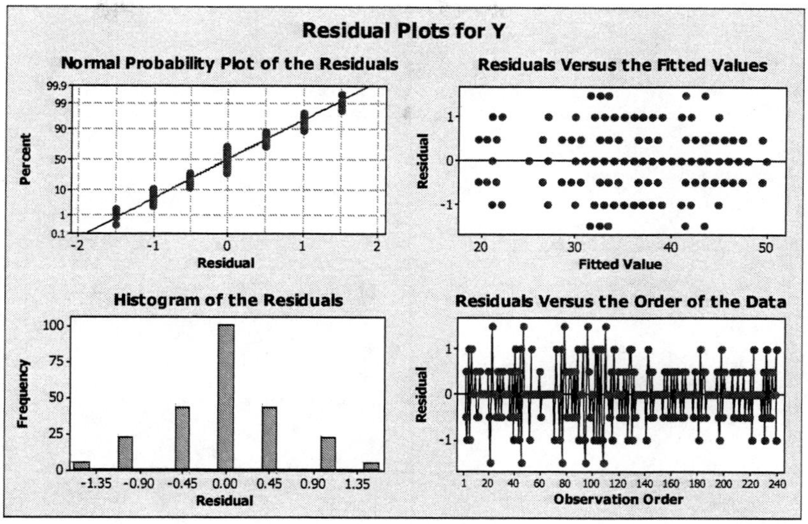

Figure 4.1. *Residual plots for the gauge R&R experiment in Table* 3.1.

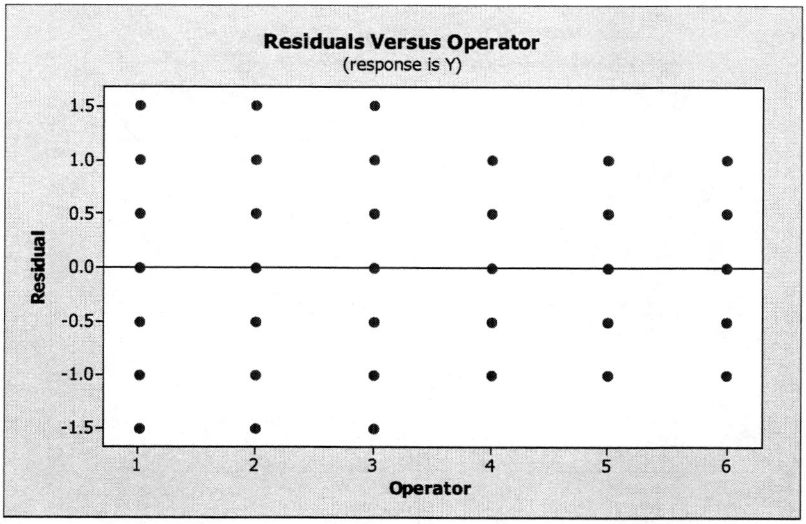

Figure 4.2. *Plot of residuals versus operators for the gauge R&R experiment in Table* 3.1.

the gauge. Out-of-control samples on the *R* chart could indicate problems with operator training, experience, or fatigue or potential differences between operators. The *R* chart in Figure 4.4 does not exhibit any unusual pattern.

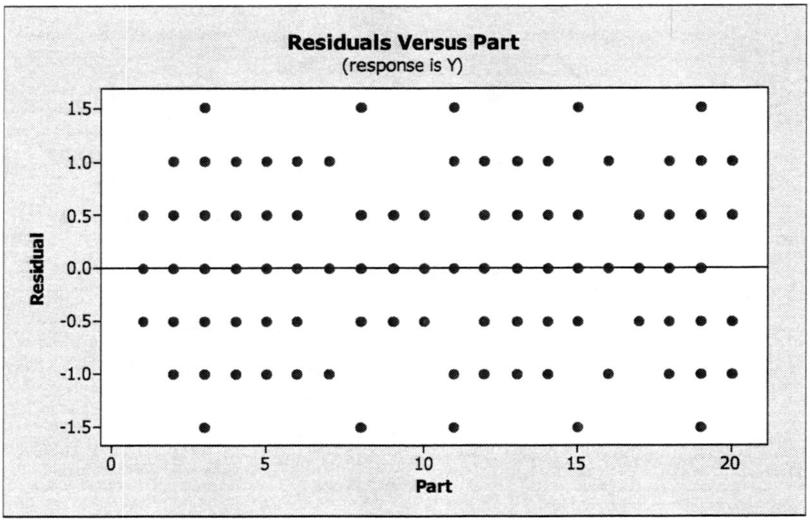

Figure 4.3. *Plot of residuals versus parts for the gauge R&R experiment in Table* 3.1.

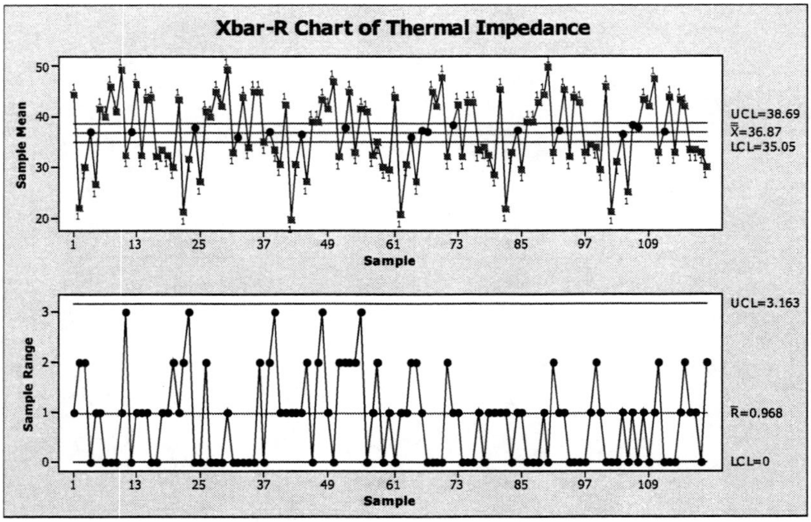

Figure 4.4. $\overline{X}$ *and R control charts for the gauge R&R experiment in Table* 3.1.

4.5 Summary

This chapter introduced some important statistical design aspects of gauge R&R experiments. A seven-step procedure useful in planning and conducting the experiment was provided. The first three steps of this procedure constitute the preexperimental planning phase and are extremely important to the successful conduct and eventual completion of the experiment. The two major types of treatment structure found in gauge R&R experiments, the factorial structure and the nested structure, were introduced and illustrated. It is also important to fully analyze the residuals from a gauge R&R experiment to investigate the adequacy of the assumed model and to determine whether there have been any substantial violations of the underlying assumptions of normality, independence of the observations, and constant variance. Residual plots were introduced as the primary means of model adequacy checking. $\overline{X}$ and R control charts are another way to examine the stability and consistency of the data from the gauge R&R experiment. The use of both residual plots and control charts was illustrated.

Chapter 5

Balanced Two-Factor Crossed Random Models with No Interaction

5.1 Introduction

As discussed in Chapter 3, interaction terms provide greater flexibility in modeling the covariance structure of the observations. However, in some situations, it is unnecessary to include interaction effects. This is particularly true if a point estimate of an interaction variance component is negative or if $H_0 : \sigma^2_{PO} = 0$ cannot be rejected using the test described in Section 3.8.1. If an investigator omits the interaction from the two-factor model in Equation (3.1), the resulting model is called the two-factor crossed random model with no interaction. This model is the subject of this chapter. We provide confidence intervals for gauge R&R parameters in Section 5.3 and for other parameters in Section 5.6. Both MLS intervals and GCIs are provided.

5.2 The Model

The balanced two-factor crossed random model with no interaction is

$$Y_{ijk} = \mu_Y + P_i + O_j + E_{ijk}, \qquad (5.1)$$
$$i = 1, \ldots, p, \quad j = 1, \ldots, o, \quad k = 1, \ldots, r,$$

where μ_Y is a constant and P_i, O_j, E_{ijk} are jointly independent normal random variables with means of zero and variances σ^2_P, σ^2_O, and σ^2_E, respectively.

The ANOVA for model (5.1) is shown in Table 5.1, and the definitions for the mean squares and means are shown in Table 5.2. Table 5.3 reports distributional properties based on the assumptions in model (5.1). Table 5.4 reports the gauge R&R parameters and point estimators. The estimators for μ_Y, γ_P, and γ_M are all MVU estimators.

5.3 MLS Intervals, Gauge R&R Parameters

Table 5.5 summarizes constants required to compute the MLS confidence intervals. The particular values reported in the table provide a confidence coefficient of 95% ($\alpha = 0.05$) for a design with $p = 100$ parts, $o = 10$ operators, and $r = 3$ replicates. These values

Table 5.1. *ANOVA for model* (5.1).

Source of variation	Degrees of freedom	Mean square	Expected mean square
Parts	$p-1$	S_P^2	$\theta_P = \sigma_E^2 + or\sigma_P^2$
Operators	$o-1$	S_O^2	$\theta_O = \sigma_E^2 + pr\sigma_O^2$
Replicates	$por-p-o+1$	S_E^2	$\theta_E = \sigma_E^2$

Table 5.2. *Mean squares and means for model* (5.1).

Statistic	Definition
S_P^2	$\dfrac{or\Sigma_i(\overline{Y}_{i**}-\overline{Y}_{***})^2}{p-1}$
S_O^2	$\dfrac{pr\Sigma_j(\overline{Y}_{*j*}-\overline{Y}_{***})^2}{o-1}$
S_E^2	$\dfrac{\Sigma_i\Sigma_j\Sigma_k(Y_{ijk}-\overline{Y}_{i**}-\overline{Y}_{*j*}+\overline{Y}_{***})^2}{por-p-o+1}$
$\overline{Y}_{i**}$	$\dfrac{\Sigma_j\Sigma_k Y_{ijk}}{or}$
$\overline{Y}_{*j*}$	$\dfrac{\Sigma_i\Sigma_k Y_{ijk}}{pr}$
$\overline{Y}_{***}$	$\dfrac{\Sigma_i\Sigma_j\Sigma_k Y_{ijk}}{por}$

Table 5.3. *Distributional results for model* (5.1).

Result	
1	$\overline{Y}_{***}$, S_P^2, S_O^2, and S_E^2 are jointly independent.
2	$(p-1)S_P^2/\theta_P$ is a chi-squared random variable with $p-1$ degrees of freedom.
3	$(o-1)S_O^2/\theta_O$ is a chi-squared random variable with $o-1$ degrees of freedom.
4	$(por-p-o+1)S_E^2/\theta_E$ is a chi-squared random variable with $por-p-o+1$ degrees of freedom.
5	$\overline{Y}_{***}$ is a normal random variable with mean μ_Y and variance $\dfrac{\theta_P+\theta_O-\theta_E}{por}$.

Table 5.4. *Gauge R&R parameters and point estimators for model* (5.1).

Gauge R&R notation	Model (5.1) representation	Point estimator
μ_Y	μ_Y	$\widehat{\mu}_Y = \overline{Y}_{***}$
γ_P	σ_P^2	$\widehat{\gamma}_P = \dfrac{S_P^2 - S_E^2}{or}$
γ_M	$\sigma_O^2 + \sigma_E^2$	$\widehat{\gamma}_M = \dfrac{S_O^2 + (pr-1)S_E^2}{pr}$
γ_R	$\dfrac{\sigma_P^2}{\sigma_O^2 + \sigma_E^2}$	$\widehat{\gamma}_R = \dfrac{\widehat{\gamma}_P}{\widehat{\gamma}_M}$

Table 5.5. *Constants used in confidence intervals for model* (5.1). *Values are for* $\alpha = 0.05$, $p = 100$, $o = 10$, *and* $r = 3$.

Constant	Definition	Value
G_1	$1 - F_{\alpha/2:\infty,p-1}$	0.2291
G_2	$1 - F_{\alpha/2:\infty,o-1}$	0.5269
G_3	$1 - F_{\alpha/2:\infty,por-p-o+1}$	0.04961
H_1	$F_{1-\alpha/2:\infty,p-1} - 1$	0.3495
H_2	$F_{1-\alpha/2:\infty,o-1} - 1$	2.333
H_3	$F_{1-\alpha/2:\infty,por-p-o+1} - 1$	0.05362
F_1	$F_{1-\alpha/2:p-1,por-p-o+1}$	1.304
F_2	$F_{\alpha/2:p-1,por-p-o+1}$	0.7380
F_3	$F_{1-\alpha/2:p-1,o-1}$	3.404
F_4	$F_{\alpha/2:p-1,o-1}$	0.4454
G_{13}	$\dfrac{(F_1-1)^2 - G_1^2F_1^2 - H_3^2}{F_1}$	0.000297
H_{13}	$\dfrac{(1-F_2)^2 - H_1^2F_2^2 - G_3^2}{F_2}$	−0.000488

correspond to the numerical example presented in Section 5.5. The reader is cautioned that some of the terms in Table 5.5 are different from what they were in Chapter 3. Make sure to reference this table when using formulas in this chapter.

5.3.1 Interval for μ_Y

A confidence interval for μ_Y under model (5.1) is obtained from the Milliken and Johnson [50] formulation used in Equation (3.2) with appropriate substitutions. This provides the approximate $100(1-\alpha)\%$ confidence interval for μ_Y,

$$L = \overline{Y}_{***} - C\sqrt{\frac{K}{por}}$$

and

$$U = \overline{Y}_{***} + C\sqrt{\frac{K}{por}}, \tag{5.2}$$

where

$$K = S_P^2 + S_O^2 - S_E^2$$

and

$$C = \frac{S_P^2\sqrt{F_{1-\alpha:1,p-1}} + S_O^2\sqrt{F_{1-\alpha:1,o-1}} - S_E^2\sqrt{F_{1-\alpha:1,por-p-o+1}}}{K}.$$

If $K < 0$, then replace K with S_E^2 and C with $\sqrt{F_{1-\alpha:1,por-p-o+1}}$.

5.3.2 Interval for γ_P

The interval for γ_P under model (5.1) is a simple modification of the interval for γ_P shown in Equation (3.3). Specifically, the approximate $100(1-\alpha)\%$ confidence interval for γ_P is

$$L = \widehat{\gamma}_P - \frac{\sqrt{V_{LP}}}{or}$$

and

$$U = \widehat{\gamma}_P + \frac{\sqrt{V_{UP}}}{or}, \tag{5.3}$$

where

$$V_{LP} = G_1^2 S_P^4 + H_3^2 S_E^4 + G_{13} S_P^2 S_E^2,$$
$$V_{UP} = H_1^2 S_P^4 + G_3^2 S_E^4 + H_{13} S_P^2 S_E^2,$$

$\widehat{\gamma}_P$ is defined in Table 5.4, and G_1, G_3, H_1, H_3, G_{13}, and H_{13} are defined in Table 5.5. Negative bounds are increased to zero.

5.3.3 Interval for γ_M

The parameter $\gamma_M = \sigma_O^2 + \sigma_E^2$ is written in terms of the expected mean squares as $[\theta_O + (pr-1)\theta_E]/(pr)$. Since γ_M is a sum of expected mean squares, we apply the MLS method proposed by Graybill and Wang [25]. The resulting bounds of an approximate $100(1-\alpha)\%$ confidence interval for γ_M are

$$L = \widehat{\gamma}_M - \frac{\sqrt{V_{LM}}}{pr}$$

and

$$U = \widehat{\gamma}_M + \frac{\sqrt{V_{UM}}}{pr}, \tag{5.4}$$

where

$$V_{LM} = G_2^2 S_O^4 + G_3^2 (pr-1)^2 S_E^4,$$
$$V_{UM} = H_2^2 S_O^4 + H_3^2 (pr-1)^2 S_E^4,$$

$\widehat{\gamma}_M$ is defined in Table 5.4, and G_2, G_3, H_2, and H_3 are defined in Table 5.5.

5.3.4 Interval for γ_R

Arteaga, Jeyaratnam, and Graybill [2] developed a confidence interval for γ_R in model (5.1). The bounds of this approximate $100(1-\alpha)\%$ confidence interval for γ_R are

$$L = \frac{p(1-G_1)S_P^4 - pS_P^2 S_E^2 + p[F_1 - (1-G_1)F_1^2]S_E^4}{o(pr-1)S_P^2 S_E^2 + o(1-G_1)F_3 S_P^2 S_O^2}$$

and

$$U = \frac{p(1+H_1)S_P^4 - pS_P^2 S_E^2 + p[F_2 - (1+H_1)F_2^2]S_E^4}{o(pr-1)S_P^2 S_E^2 + o(1+H_1)F_4 S_P^2 S_O^2}, \qquad (5.5)$$

where G_1, H_1, F_1, F_2, F_3, and F_4 are defined in Table 5.5. Negative bounds are increased to zero.

5.4 Generalized Intervals, Gauge R&R Parameters

Table 5.6 reports generalized pivotal quantities for the gauge R&R parameters in model (5.1), where Z is a standard normal random variable and W_1, W_2, and W_3 are jointly independent chi-squared random variables that are independent of Z with degrees of freedom, $p-1$, $o-1$, and $por-p-o+1$, respectively. The terms s_P^2, s_O^2, and s_E^2 are the realized values of the mean squares.

The process used to compute GCIs is the same one described in Section 3.4 with obvious substitutions. The confidence intervals for the misclassification rates are computed in the manner described in Section 3.4 using the GPQs shown in Table 5.7.

Table 5.6. *GPQs for gauge R&R parameters in model* (5.1).

Parameter	GPQ
μ_Y	$\bar{y}_{***} - Z\sqrt{\max\left[\epsilon, \frac{(p-1)s_P^2}{por W_1} + \frac{(o-1)s_O^2}{por W_2} - \frac{(por-p-o+1)s_E^2}{por W_3}\right]}$
γ_P	$\max\left[0, \frac{(p-1)s_P^2}{or W_1} - \frac{(por-p-o+1)s_E^2}{or W_3}\right]$
γ_M	$\frac{(o-1)s_O^2}{pr W_2} + \frac{(pr-1)(por-p-o+1)s_E^2}{pr W_3}$
γ_R	$\frac{\text{GPQ}(\gamma_P)}{\text{GPQ}(\gamma_M)}$

Table 5.7. *GPQs for misclassification rates in model* (5.1).

Parameter	GPQ
$\mu_Y = \mu_P$	$\overline{y}_{***} - Z\sqrt{\max\left[\epsilon, \frac{(p-1)s_P^2}{porW_1} + \frac{(o-1)s_O^2}{porW_2} - \frac{(por-p-o+1)s_E^2}{porW_3}\right]}$
$\gamma_P + \gamma_M$	$\frac{(p-1)s_P^2}{orW_1} + \frac{(o-1)s_O^2}{prW_2} + \frac{(por-p-o)(por-p-o+1)s_E^2}{porW_3}$
γ_P	$\max\left[\epsilon, \frac{(p-1)s_P^2}{orW_1} - \frac{(por-p-o+1)s_E^2}{orW_3}\right]$

5.5 Numerical Example

We now use the formulas presented in this section to analyze an experiment concerning production of tape drive heads. The head is the part of a tape drive that makes contact with the tape and magnetically records information on the tape. A new head tester has been developed, and it must be determined if it is capable of monitoring the manufacturing process. In this study, 100 heads (parts) are selected at random from the manufacturing process that produces the heads. Then 10 different tape cartridges (operators) are selected at random to record measurements in the head tester. There are three replicates for each head × tape combination. The response variable is reverse resolution. In the process of writing information to magnetic tape, voltage is measured at two different frequencies. Reverse resolution is the ratio of these two voltages when the drive writes in the reverse direction. There is no unit of measurement on the ratio, and it is expressed as a percentage. The specification limits are $LSL = 90\%$ and $USL = 110\%$. Table 5.8 shows a subset of the data. No interaction between heads and tapes is expected, and so model (5.1) is selected for the analysis. The computed ANOVA for the entire data set is shown in Table 5.9.

To construct 95% confidence intervals, we use the constants shown in Table 5.5 with $\alpha = 0.05$, $p = 100$ heads, $o = 10$ tapes, and $r = 3$ replicates. The mean of all the observations is 100.2.

5.5.1 Interval for μ_Y

The point estimate for μ_Y is the sample mean 100.2. The confidence interval is shown in Equation (5.2). Here, $K = S_P^2 + S_O^2 - S_E^2 = 307.9$ and $C = 1.998$ using $F_{1-\alpha:1,p-1} = F_{0.95:1,99} = 3.937$, $F_{1-\alpha:1,o-1} = F_{0.95:1,9} = 5.117$, and $F_{1-\alpha:1,por-p-o+1} = F_{0.95:1,2891} = 3.845$. Hence the 95% confidence interval for μ_Y is

$$\begin{aligned} L &= \overline{Y}_{***} - C\sqrt{\frac{K}{por}} \\ &= 100.2 - 1.998\sqrt{\frac{307.9}{3{,}000}} \\ &= 99.6 \end{aligned}$$

Table 5.8. *Partial data for tape head study.*

	Tape			
Head	1	2	...	10
1	104.235	103.677	...	104.324
	104.084	103.538	...	104.476
	103.767	103.727	...	104.451
2	93.773	93.117	...	93.771
	93.510	93.217	...	93.917
	93.932	93.408	...	93.766
⋮	⋮	⋮	...	⋮
100	99.166	98.758	...	99.046
	98.878	99.071	...	98.946
	98.659	98.514	...	99.081

Table 5.9. *ANOVA for tape head example.*

Source of variation	Degrees of freedom	Mean square
Heads	99	292.7
Tapes	9	15.26
Replicates	2891	0.05159

and

$$\begin{aligned} U &= \overline{Y}_{***} + C\sqrt{\frac{K}{por}} \\ &= 100.2 + 1.998\sqrt{\frac{307.9}{3{,}000}} \\ &= 101. \end{aligned}$$

The GPQ for μ_Y from Table 5.6 is

$$\begin{aligned} \text{GPQ}(\mu_Y) &= \overline{y}_{***} - Z\sqrt{\max\left[\epsilon, \frac{(p-1)s_P^2}{por\,W_1} + \frac{(o-1)s_O^2}{por\,W_2} - \frac{(por-p-o+1)s_E^2}{por\,W_3}\right]} \\ &= 100.2 - Z\sqrt{\max\left[0.001, \frac{99(292.7)}{3{,}000W_1} + \frac{9(15.26)}{3{,}000W_2} - \frac{2891(0.05159)}{3{,}000W_3}\right]}, \end{aligned}$$

where Z is a standard normal random variable, W_1, W_2, and W_3 are jointly independent chi-squared random variables that are independent of Z with degrees of freedom 99, 9, and 2,891, respectively, and $\epsilon = 0.001$. The 95% GCI based on 100,000 simulated GPQ values is from 99.6 to 101.

5.5.2 Interval for γ_P

The point estimate for γ_P is

$$\begin{aligned}\widehat{\gamma}_P &= \frac{S_P^2 - S_E^2}{or} \\ &= \frac{292.7 - 0.05159}{30} \\ &= 9.755.\end{aligned}$$

The lower and upper confidence bounds are shown in Equation (5.3). In this example,

$$\begin{aligned}V_{LP} &= G_1^2 S_P^4 + H_3^2 S_E^4 + G_{13} S_P^2 S_E^2 \\ &= (0.2291)^2(292.7)^2 + (0.05362)^2(0.05159)^2 + (0.000297)(292.7)(0.05159) \\ &= 4{,}497\end{aligned}$$

and

$$\begin{aligned}V_{UP} &= H_1^2 S_P^4 + G_3^2 S_E^4 + H_{13} S_P^2 S_E^2 \\ &= (0.3495)^2(292.7)^2 + (0.04961)^2(0.05159)^2 + (-0.000488)(292.7)(0.05159) \\ &= 10{,}465.\end{aligned}$$

Substituting this information into Equation (5.3) yields the 95% confidence interval

$$L = 9.755 - \frac{\sqrt{4{,}497}}{30} = 7.52$$

and

$$U = 9.755 + \frac{\sqrt{10{,}465}}{30} = 13.2. \tag{5.6}$$

The GPQ for γ_P is shown in Table 5.6. For the given data set, this GPQ is

$$\begin{aligned}\text{GPQ}(\gamma_P) &= \max\left[0, \frac{(p-1)s_P^2}{or\,W_1} - \frac{(por - p - o + 1)s_E^2}{or\,W_3}\right] \\ &= \max\left[0, \frac{99(292.7)}{30W_1} - \frac{2891(0.05159)}{30W_3}\right],\end{aligned}$$

where W_1 and W_3 are independent chi-squared random variables with 99 and 2,891 degrees of freedom, respectively. The 95% generalized interval for γ_P based on 100,000 simulated values of W_1 and W_3 is from $L = 7.52$ to $U = 13.2$.

5.5.3 Interval for γ_M

The point estimate for γ_M is

$$\begin{aligned}\widehat{\gamma}_M &= \frac{S_O^2 + (pr-1)S_E^2}{pr} \\ &= \frac{15.26 + 299(0.05159)}{300} \\ &= 0.102.\end{aligned}$$

The lower and upper confidence bounds are shown in Equation (5.4). In this example,

$$\begin{aligned}V_{LM} &= G_2^2 S_O^4 + G_3^2(pr-1)^2 S_E^4 \\ &= (0.5269)^2(15.26)^2 + (0.04961)^2(299)^2(0.05159)^2 \\ &= 65.23\end{aligned}$$

and

$$\begin{aligned}V_{UM} &= H_2^2 S_O^4 + H_3^2(pr-1)^2 S_E^4 \\ &= (2.333)^2(15.26)^2 + (0.05362)^2(299)^2(0.05159)^2 \\ &= 1{,}268.\end{aligned}$$

Substituting this information into Equation (5.4) yields the 95% confidence interval,

$$L = 0.102 - \frac{\sqrt{65.23}}{300} = 0.075$$

and

$$U = 0.102 + \frac{\sqrt{1{,}268}}{300} = 0.221. \tag{5.7}$$

The GPQ for γ_M shown in Table 5.6 is

$$\begin{aligned}\mathrm{GPQ}(\gamma_M) &= \frac{(o-1)s_O^2}{pr\,W_2} + \frac{(pr-1)(por-p-o+1)s_E^2}{pr\,W_3} \\ &= \frac{9(15.26)}{300W_2} + \frac{299(2891)(0.05159)}{300W_3},\end{aligned}$$

where W_2 and W_3 are independent chi-squared random variables with 9 and 2,891 degrees of freedom, respectively. Our computed 95% generalized interval for γ_M is from $L = 0.075$ to $U = 0.222$.

5.5.4 Interval for γ_R

The point estimate for γ_R is

$$\begin{aligned}\widehat{\gamma}_R &= \frac{\widehat{\gamma}_P}{\widehat{\gamma}_M} \\ &= \frac{9.755}{0.102} \\ &= 95.6.\end{aligned}$$

Using Equation (5.5) to compute a 95% confidence interval for γ_R, we obtain

$$\begin{aligned}
L &= \frac{p}{o} \times \frac{(1-G_1)S_P^4 - S_P^2 S_E^2 + [F_1 - (1-G_1)F_1^2]S_E^4}{(pr-1)S_P^2 S_E^2 + (1-G_1)F_3 S_P^2 S_O^2} \\
&= \frac{100}{10} \times \frac{0.7709(292.7)^2 - 292.7(0.05159) + [1.304 - (0.7709)(1.304)^2](0.05159)^2}{299(292.7)(0.05159) + 0.7709(3.404)(292.7)(15.26)} \\
&= 40.7
\end{aligned}$$

and

$$\begin{aligned}
U &= \frac{p}{o} \times \frac{(1+H_1)S_P^4 - S_P^2 S_E^2 + [F_2 - (1+H_1)F_2^2]S_E^4}{(pr-1)S_P^2 S_E^2 + (1+H_1)F_4 S_P^2 S_O^2} \\
&= \frac{100}{10} \times \frac{1.349(292.7)^2 - 292.7(0.05159) + [0.7380 - (1.349)(0.7380)^2](0.05159)^2}{299(292.7)(0.05159) + 1.349(0.4454)(292.7)(15.26)} \\
&= 161. \qquad (5.8)
\end{aligned}$$

The 95% GCI formed using the GPQ for γ_R shown in Table 5.6 is from $L = 42.3$ to $U = 146$.

5.5.5 Interval for PTR

Using Equation (1.2) with $k = 6$, a point estimate for PTR is

$$\frac{k\sqrt{\hat{\gamma}_M}}{USL - LSL} = \frac{6\sqrt{0.102}}{110 - 90} = 0.096.$$

A 95% confidence interval for PTR based on the bounds calculated in Equation (5.7) is

$$L = \frac{6\sqrt{0.075}}{110 - 90} = 0.082$$

and

$$U = \frac{6\sqrt{0.221}}{110 - 90} = 0.141.$$

5.5.6 Interval for SNR

Using Equation (1.3), the point estimate for SNR is

$$\sqrt{2\hat{\gamma}_R} = \sqrt{2(95.6)} = 13.8.$$

A 95% confidence interval for SNR based on the bounds for γ_R calculated in Equation (5.8) is

$$L = \sqrt{2(40.7)} = 9.02$$

and

$$U = \sqrt{2(161)} = 17.9.$$

Table 5.10. *95% confidence intervals for misclassification rates. (Bounds for δ, δ_c, β, and β_c have been multiplied by 10^6.)*

Parameter	Lower bound	Upper bound
δ	70	871
δ_c	70	877
δ_{index}	0.120	0.313
β	41	611
β_c	88,974	163,983
β_{index}	0.089	0.164

5.5.7 Interval for C_p

Using the relation in Equation (1.5), a point estimate for C_p is

$$\frac{110-90}{6\sqrt{9.755}} = 1.07.$$

The 95% confidence interval for C_p based on the confidence interval in Equation (5.6) is

$$L = \frac{110-90}{6\sqrt{13.2}} = 0.92$$

and

$$U = \frac{110-90}{6\sqrt{7.52}} = 1.2.$$

5.5.8 Intervals for Misclassification Rates

The 95% confidence intervals for the misclassification rates based on the algorithm in Section 3.4 is shown in Table 5.10. Since the upper bounds for δ_{index} and β_{index} are both less than one, this suggests the measurement system is able to discriminate parts. This result is consistent with the confidence intervals for PTR and SNR.

5.5.9 Conclusions

Table 5.11 reports the MLS intervals and GCIs for our example. As shown in the table, there is not much practical difference between the two sets of intervals.

The R&R graph is shown in Figure 5.1. The dashed rectangle formed by the intersection of the confidence intervals for PTR and SNR is located in Region 1. This implies that the measurement system satisfies both the PTR and SNR criteria.

Figure 5.2 reports the residual plots described in Chapter 4. These plots suggest the required assumptions are reasonable.

Table 5.11. *Summary of 95% confidence intervals for example. (See Section 1.8 for a description of computer programs to perform these computations.)*

Parameter	MLS	GCI
μ_Y	$L = 99.6$ $U = 101$	$L = 99.6$ $U = 101$
γ_P	$L = 7.52$ $U = 13.2$	$L = 7.52$ $U = 13.2$
γ_M	$L = 0.075$ $U = 0.221$	$L = 0.075$ $U = 0.222$
γ_R	$L = 40.7$ $U = 161$	$L = 42.3$ $U = 146$
PTR	$L = 0.082$ $U = 0.141$	$L = 0.082$ $U = 0.141$
SNR	$L = 9.02$ $U = 17.9$	$L = 9.20$ $U = 17.1$
C_p	$L = 0.92$ $U = 1.2$	$L = 0.92$ $U = 1.2$

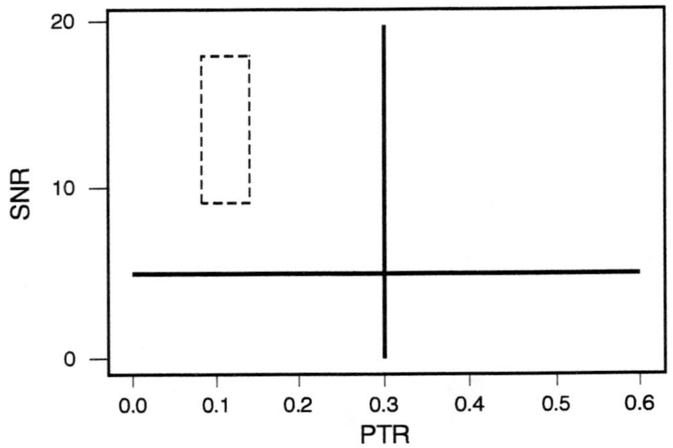

Figure 5.1. *R&R graph for two-factor with no interaction example.*

5.6 Intervals for Other Parameters

In this section, we provide brief results for parameters that are sometimes of interest in applications other than gauge R&R studies. Table 5.12 reports constants not defined in Table 5.5 that are used in the MLS intervals that follow.

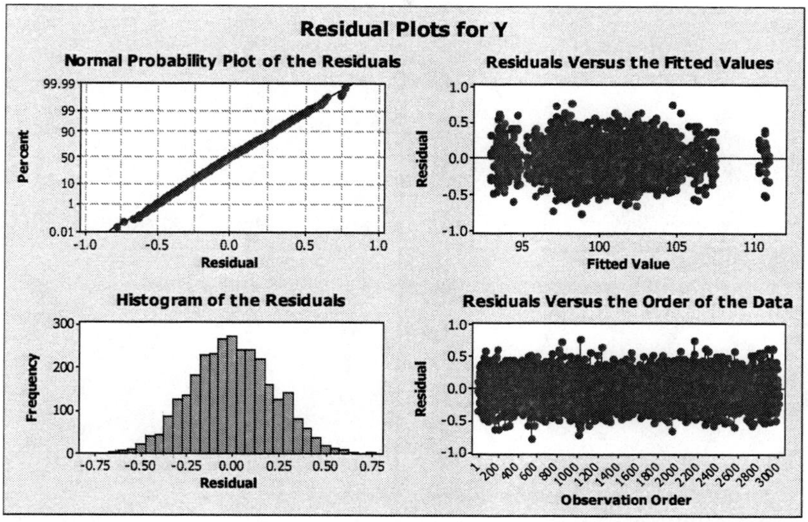

Figure 5.2. *Residual plots for the two-factor model with no interaction.*

Table 5.12. *Additional constants used in Section* 5.6. *Values are for* $\alpha = 0.05$, $p = 100$, $o = 10$, *and* $r = 3$.

Constant	Definition	Value
F_5	$F_{1-\alpha/2:o-1,por-p-o+1}$	2.118
F_6	$F_{\alpha/2:o-1,por-p-o+1}$	0.2998
G_{23}	$\dfrac{(F_5-1)^2 - G_2^2F_5^2 - H_3^2}{F_5}$	0.000838
H_{23}	$\dfrac{(1-F_6)^2 - H_2^2F_6^2 - G_3^2}{F_6}$	−0.00474

5.6.1 Intervals for σ_O^2 and σ_E^2

A confidence interval for $\sigma_O^2 = (\theta_O - \theta_E)/(pr)$ is computed using the method described in Section 5.3.2 with appropriate substitutions. In particular, the bounds of an approximate $100(1-\alpha)\%$ confidence interval for σ_O^2 are

$$L = \widehat{\sigma}_O^2 - \frac{\sqrt{V_{LO}}}{pr}$$

and

$$U = \widehat{\sigma}_O^2 + \frac{\sqrt{V_{UO}}}{pr}, \tag{5.9}$$

where

$$\widehat{\sigma}_O^2 = \frac{S_O^2 - S_E^2}{pr},$$

$$V_{LO} = G_2^2 S_O^4 + H_3^2 S_E^4 + G_{23} S_O^2 S_E^2,$$

$$V_{UO} = H_2^2 S_O^4 + G_3^2 S_E^4 + H_{23} S_O^2 S_E^2,$$

G_2, G_3, H_2, and H_3 are defined in Table 5.5, and G_{23} and H_{23} are defined in Table 5.12. Negative bounds are increased to zero.

The exact $100(1-\alpha)\%$ confidence interval for σ_E^2 using Result 4 in Table 5.3 is

$$L = (1 - G_3) S_E^2$$

and

$$U = (1 + H_3) S_E^2, \tag{5.10}$$

where G_3 and H_3 are defined in Table 5.5.

5.6.2 Interval for γ_Y

The total variation of the response variable is $\gamma_Y = \sigma_P^2 + \sigma_O^2 + \sigma_E^2 = [p\theta_P + o\theta_O + (por - p - o)\theta_E]/(por)$. Since this is a sum of expected mean squares, we apply the Graybill and Wang [25] method to construct the interval. The resulting bounds of an approximate $100(1-\alpha)\%$ confidence interval are

$$L = \widehat{\gamma}_Y - \frac{\sqrt{V_{LT}}}{por}$$

and

$$U = \widehat{\gamma}_Y + \frac{\sqrt{V_{UT}}}{por},$$

where

$$\widehat{\gamma}_Y = \frac{pS_P^2 + oS_O^2 + (por - p - o)S_E^2}{por},$$

$$V_{LT} = G_1^2 p^2 S_P^4 + G_2^2 o^2 S_O^4 + G_3^2 (por - p - o)^2 S_E^4,$$

$$V_{UT} = H_1^2 p^2 S_P^4 + H_2^2 o^2 S_O^4 + H_3^2 (por - p - o)^2 S_E^4,$$

and G_1, G_2, G_3, H_1, H_2, and H_3 are defined in Table 5.5.

5.6.3 Intervals for σ_P^2/σ_E^2 and σ_O^2/σ_E^2

Exact confidence intervals can be computed for σ_P^2/σ_E^2 and σ_O^2/σ_E^2 by noting that $(S_P^2\theta_E)/(S_E^2\theta_P)$ and $(S_O^2\theta_E)/(S_E^2\theta_O)$ have exact F-distributions. This is a consequence of Results 1 through 4 in Table 5.3. This provides the exact $100(1-\alpha)\%$ confidence interval for σ_P^2/σ_E^2,

$$L = \frac{1}{or}\left[\frac{S_P^2}{S_E^2 F_1} - 1\right]$$

and

$$U = \frac{1}{or}\left[\frac{S_P^2}{S_E^2 F_2} - 1\right], \tag{5.11}$$

and the exact $100(1-\alpha)\%$ confidence interval for σ_O^2/σ_E^2,

$$L = \frac{1}{pr}\left[\frac{S_O^2}{S_E^2 F_5} - 1\right]$$

and

$$U = \frac{1}{pr}\left[\frac{S_O^2}{S_E^2 F_6} - 1\right], \tag{5.12}$$

where F_1 and F_2 are defined in Table 5.5 and F_5 and F_6 are defined in Table 5.12.

Exact intervals for $\sigma_P^2/(\sigma_P^2+\sigma_E^2)$, $\sigma_O^2/(\sigma_O^2+\sigma_E^2)$, $\sigma_E^2/(\sigma_P^2+\sigma_E^2)$, and $\sigma_E^2/(\sigma_O^2+\sigma_E^2)$ are all obtained from transformations of Equations (5.11) and (5.12). For example, if L and U represent the interval for σ_P^2/σ_E^2 in Equation (5.11), then a $100(1-\alpha)\%$ confidence interval for $\sigma_P^2/(\sigma_P^2+\sigma_E^2)$ has lower bound $L/(1+L)$ and upper bound $U/(1+U)$.

5.6.4 Intervals for σ_O^2/γ_Y and σ_E^2/γ_Y

A confidence interval for σ_O^2/γ_Y is based on a modification of the interval for γ_R in Equation (5.5). We begin with the $100(1-\alpha)\%$ confidence interval for $\sigma_O^2/(\sigma_P^2+\sigma_E^2)$,

$$L^* = \frac{o(1-G_2)S_O^4 - oS_O^2S_E^2 + o[F_5-(1-G_2)F_5^2]S_E^4}{p(or-1)S_O^2S_E^2 + p(1-G_2)S_O^2S_P^2/F_4}$$

and

$$U^* = \frac{o(1+H_2)S_O^4 - oS_O^2S_E^2 + o[F_6-(1+H_2)F_6^2]S_E^4}{p(or-1)S_O^2S_E^2 + p(1+H_2)S_O^2S_P^2/F_3}, \tag{5.13}$$

where G_2, H_2, F_3, and F_4 are defined in Table 5.5 and F_5 and F_6 are defined in Table 5.12. The $100(1-\alpha)\%$ confidence interval for σ_O^2/γ_Y is then

$$L = \frac{L^*}{1+L^*}$$

and

$$U = \frac{U^*}{1+U^*}, \tag{5.14}$$

where L^* and U^* are defined in Equation (5.13) and set to zero if they are negative. The MLS confidence interval for σ_E^2/γ_Y can be computed using the general results described in [10, p. 130].

Generalized pivotal quantities for the parameters in this section are shown in Table 5.13, where W_1, W_2, and W_3 are jointly independent chi-squared random variables with degrees of freedom $p-1$, $o-1$, and $por-p-o+1$, respectively. The terms s_P^2, s_O^2, and s_E^2 are the realized values of the mean squares. Since the intervals in Equations (5.10), (5.11), and (5.12) are exact, no GPQs are reported for σ_E^2, σ_P^2/σ_E^2, or σ_O^2/σ_E^2. Table 5.14 reports the computed intervals for the example data presented in Section 5.5.

Table 5.13. *GPQs for other parameters in model* (5.1).

Parameter	GPQ
σ_O^2	$\max\left[0, \dfrac{(o-1)s_O^2}{prW_2} - \dfrac{(por-p-o+1)s_E^2}{prW_3}\right]$
γ_Y	$\dfrac{(p-1)s_P^2}{orW_1} + \dfrac{(o-1)s_O^2}{prW_2} + \dfrac{(por-p-o)(por-p-o+1)s_E^2}{porW_3}$
σ_O^2/γ_Y	$\dfrac{\text{GPQ}(\sigma_O^2)}{\text{GPQ}(\gamma_Y)}$
σ_E^2/γ_Y	$\dfrac{(por-p-o+1)s_E^2/W_3}{\text{GPQ}(\gamma_Y)}$

Table 5.14. *Computed intervals for other parameters in example. (See Section* 1.8 *for a description of computer programs to perform these computations.)*

Parameter	MLS	GCI
σ_O^2	$L = 0.0239$ $U = 0.169$	$L = 0.0240$ $U = 0.169$
σ_E^2	$L = 0.0490$ $U = 0.0544$	Exact
γ_Y	$L = 7.62$ $U = 13.3$	$L = 7.64$ $U = 13.3$
σ_P^2/σ_E^2	$L = 145$ $U = 256$	Exact
σ_O^2/σ_E^2	$L = 0.462$ $U = 3.29$	Exact
σ_O^2/γ_Y	$L = 0.00229$ $U = 0.0173$	$L = 0.00230$ $U = 0.0173$
σ_E^2/γ_Y		$L = 0.00387$ $U = 0.0068$

5.7 Summary

In this chapter we presented results for a random two-factor model with no interaction. An investigator should use the formulas in this chapter instead of those in Chapter 3 when there is no evidence of interaction. In the next chapter we consider both models when operators are fixed. Intervals can be obtained for this mixed model either by using generalized inference or by modifying the MLS intervals given in Chapters 3 and 5.

Chapter 6

Balanced Two-Factor Crossed Mixed Models

6.1 Introduction

In Chapters 3 and 5 we presented two-factor models where both factors are assumed to be random effects. Although this has traditionally been the assumption in gauge R&R studies, it is often more appropriate to treat operators as fixed effects. Consider the situation described by Dolezal, Burdick, and Birch [18], in which three mechanical testers (operators) are used to monitor a process that manufactures tape drive heads (parts). A random sample of 18 heads is obtained from the process output, and the response variable "reverse overwrite" was measured for each head using the set of three testers. Reverse overwrite is a measure of residual frequency after a second frequency is placed on magnetic tape. The units of measurement are decibels and the specification limits are $LSL = -41$ and $USL = -33$. Each tester makes three replicate measurements on each head. The total data set consists of 162 measurements. A partial listing of the data is given in Table 6.1. The testers in this situation are each attached to one of three production lines and will always be part of the testing system. Since these are the only three testers that will ever be used to monitor the process, the inference concerns only these three testers and not the population from which they were selected. Thus, the operator factor is a fixed effect. (If you are not familiar with the terms "random" and "fixed" in this context, review the material in Appendix A.) The two-factor mixed model with random parts and fixed operators is the topic of this chapter.

6.2 The Mixed Model Formulation

For a set of o fixed operators, the measurements from the jth operator are represented as

$$Y_j = X + E_j, \tag{6.1}$$

where Y_j is a measured value of a randomly selected part, X is the true value, and E_j is the measurement error. The terms X and E_j are independent normal random variables with means μ_P and μ_{Ej} and variances γ_P and γ_E, respectively. These assumptions imply that the o populations of operator measurements have equal variances but possibly different means. In particular, the population of measurements for the jth operator is normal with mean

Table 6.1. *Responses from head tester experiment (in decibels).*

Part	Operator 1	Operator 2	Operator 3
1	−40.9928	−40.8646	−40.4210
	−40.9907	−40.8775	−40.4155
	−40.9852	−40.8732	−40.5255
2	−39.8866	−40.0948	−39.9957
	−39.8904	−39.9804	−39.9976
	−39.6693	−39.7833	−39.6829
...	...	...	...
17	−40.6839	−41.6339	−40.7640
	−39.9705	−40.6301	−40.0153
	−40.1715	−40.3670	−40.2802
18	−39.0619	−39.8868	−39.1646
	−38.9755	−39.3642	−39.0721
	−39.0646	−39.5285	−39.1462

Table 6.2. *Probability distribution of J with three operators.*

j	$E(Y_j \mid J = j)$	Probability
1	μ_1	1/3
2	μ_2	1/3
3	μ_3	1/3

$\mu_j = \mu_P + \mu_{Ej}$ and variance $\gamma_P + \gamma_E$. As discussed in Section 1.2, we assume there is no systematic bias in the measurement system. This implies that $\sum_j \mu_{Ej} = 0$ so that

$$\begin{aligned}\mu_Y &= \frac{\sum_j \mu_j}{o} \\ &= \frac{\sum_j (\mu_P + \mu_{Ej})}{o} \\ &= \mu_P.\end{aligned}$$

As before, this is an assumption of convenience since it does not affect estimation of the variances.

When defining total variance of the measurement system, it is necessary to consider variation in the combined measurements from all operators. To demonstrate, consider a random variable J that assigns value j with probability $1/o$ for $j = 1, 2, \ldots, o$. This random variable represents the operator making a randomly selected measurement. Based on model (6.1), we have stated that $E(Y_j|J = j) = \mu_j$ and $V(Y_j|J = j) = \gamma_P + \gamma_E$, where E and V are the expected value and variance operators, respectively. The probability distribution for J with $o = 3$ operators is shown in Table 6.2. We denote the total variance

Table 6.3. *Gauge R&R parameters in mixed model with fixed operators.*

Symbol	Definition
$\mu_Y = \frac{\sum_j \mu_j}{o}$	Mean of population of measurements from all operators
μ_j	Mean of population of measurements from operator j
γ_P	Variance of the process
$\gamma_M = \gamma_O + \gamma_E$	Variance of the measurement system
$\gamma_R = \frac{\gamma_P}{\gamma_M}$	Ratio of process variance to measurement variance

in the combined measurements of all operators as $V(Y)$, where

$$V(Y) = V[E(Y_j|J)] + E[V(Y_j|J)].$$

(See, e.g., Lohr [44, pp. 433–435].) Since $V(Y_j|J)$ is constant for all values of J, then $E[V(Y_j|J)] = \gamma_P + \gamma_E$. Now

$$\begin{aligned} V[E(Y_j|J)] &= \frac{\sum_j \mu_j^2}{o} - \{E[E(Y_j|J)]\}^2 \\ &= \frac{\sum_j \mu_j^2}{o} - \mu_Y^2 \\ &= \frac{\sum_j (\mu_j - \mu_Y)^2}{o}. \end{aligned}$$

So the total variance in the combined measurements is

$$\begin{aligned} \gamma_Y &= \gamma_O + \gamma_P + \gamma_E \\ &= \gamma_P + \gamma_M, \end{aligned}$$

where

$$\gamma_O = \frac{\sum_j (\mu_j - \mu_Y)^2}{o} \tag{6.2}$$

and

$$\gamma_M = \gamma_O + \gamma_E.$$

Using these definitions, the parameters in a gauge R&R study with fixed operators are as shown in Table 6.3.

Note that if each operator is centered at μ_P, then $\mu_j - \mu_Y = 0$ for all j and $\gamma_O = 0$. In this situation, there is no operator effect and the data can be analyzed using a random effects model. For example, if there is no evidence of an operator effect for the data in Table 6.1 (i.e., if there is no evidence that $\gamma_O > 0$), then the data can be analyzed using the one-factor random effects model described in Chapter 2 with $p = 18$ parts and $r = 9$ replications.

Table 6.4. *ANOVA for model* (6.3).

Source of variation	Degrees of freedom	Mean square	Expected mean square
Parts (P)	$p-1$	S_P^2	$\theta_P = \sigma_E^2 + r\sigma_{PO}^2 + or\sigma_P^2$
Operators (O)	$o-1$	S_O^2	$\theta_O = \sigma_E^2 + r\sigma_{PO}^2 + \left(\frac{pro}{o-1}\right)\gamma_o$
P×O	$(p-1)(o-1)$	S_{PO}^2	$\theta_{PO} = \sigma_E^2 + r\sigma_{PO}^2$
Replicates	$po(r-1)$	S_E^2	$\theta_E = \sigma_E^2$

6.3 The Two-Factor Mixed Model with Interaction

The balanced two-factor mixed model with interaction is

$$Y_{ijk} = P_i + \mu_j + (PO)_{ij} + E_{ijk}, \qquad (6.3)$$
$$i = 1, \ldots, p, \quad j = 1, \ldots, o, \quad k = 1, \ldots, r,$$

where μ_j is a constant specific to the jth operator, and P_i, $(PO)_{ij}$, and E_{ijk} are jointly independent normal random variables with means of zero and variances σ_P^2, σ_{PO}^2, and σ_E^2, respectively. In the context of a gauge R&R study, parts are random effects and operators are fixed effects. We place no restrictions on the $(PO)_{ij}$ since we assume that measurements on the same part by different operators will be positively related. (See Section 7.2 of Burdick and Graybill [10] and references cited therein for a discussion of restricted and unrestricted formulations of the mixed model.)

The ANOVA for model (6.3) is shown in Table 6.4, where γ_o is defined in Equation (6.2). This table is identical to Table 3.2 except that $o\gamma_o/(o-1)$ replaces σ_O^2 in θ_O. The definitions for the mean squares and means are the same ones shown in Table 3.3. For convenience these are repeated in Table 6.5. Table 6.6 reports distributional properties based on the assumptions in model (6.3), and Table 6.7 presents the resulting covariance structure. This is the same structure shown in Table 3.5 with $\sigma_O^2 = 0$. Table 6.8 reports the gauge R&R parameters and point estimators. These are the parameters presented in Table 6.3 written in the notation of model (6.3).

Note from Result 3 in Table 6.6 that Chapter 3 intervals containing S_O^2 are no longer valid since $(o-1)S_O^2/\theta_O$ no longer has a chi-squared distribution. We present methods to handle this problem in the next two sections. The first approach modifies the MLS intervals reported in Chapter 3. The second approach employs GCIs.

6.4 MLS Intervals, Gauge R&R Parameters

Dolezal, Burdick, and Birch [18] proposed a method for constructing confidence intervals in model (6.3) by modifying the intervals reported in Chapter 3. The basic idea is to approximate the distribution of S_O^2 with a chi-squared distribution and then use the Chapter 3 formulas with this approximation.

Table 6.5. *Mean squares and means for model* (6.3).

Statistic	Definition
S_P^2	$\dfrac{or\,\Sigma_i(\overline{Y}_{i**} - \overline{Y}_{***})^2}{p-1}$
S_O^2	$\dfrac{pr\,\Sigma_j(\overline{Y}_{*j*} - \overline{Y}_{***})^2}{o-1}$
S_{PO}^2	$\dfrac{r\,\Sigma_i\Sigma_j(\overline{Y}_{ij*} - \overline{Y}_{i**} - \overline{Y}_{*j*} + \overline{Y}_{***})^2}{(p-1)(o-1)}$
S_E^2	$\dfrac{\Sigma_i\Sigma_j\Sigma_k(Y_{ijk} - \overline{Y}_{ij*})^2}{po(r-1)}$
$\overline{Y}_{i**}$	$\dfrac{\Sigma_j\Sigma_k Y_{ijk}}{or}$
$\overline{Y}_{*j*}$	$\dfrac{\Sigma_i\Sigma_k Y_{ijk}}{pr}$
$\overline{Y}_{ij*}$	$\dfrac{\Sigma_k Y_{ijk}}{r}$
$\overline{Y}_{***}$	$\dfrac{\Sigma_i\Sigma_j\Sigma_k Y_{ijk}}{por}$

In particular, $n^* S_O^2/\theta_O$ is approximated as a chi-squared random variable with n^* degrees of freedom, where

$$n^* = \frac{[(o-1) + 2\widehat{\lambda}]^2}{(o-1) + 4\widehat{\lambda}}$$

and

$$\widehat{\lambda} = \frac{o-1}{2}\left[\frac{S_O^2}{S_{PO}^2}\left\{\frac{(p-1)(o-1)-2}{(p-1)(o-1)}\right\} - 1\right] \tag{6.4}$$

for $(p-1)(o-1) > 2$. If $\widehat{\lambda}$ is negative, $n^* = o - 1$. Throughout the book, we truncate n^* to the greatest integer less than or equal to n^*.

Now to form a confidence interval, one simply uses n^* as the operator degrees of freedom in the Chapter 3 formulas. Also, γ_o replaces σ_O^2 in the definitions of γ_M and γ_R. We now present the MLS confidence intervals for the gauge R&R parameters using these substitutions. Table 6.9 summarizes constants required to compute the intervals. The particular values reported in Table 6.9 provide a two-sided interval with 95% confidence ($\alpha = 0.05$) for a design with $p = 18$ parts, $o = 3$ operators, $r = 3$ replicates, and $n^* = 25$.

Table 6.6. *Distributional results for model* (6.3).

Result	
1	$\overline{Y}_{*j*}$, S_P^2, S_O^2, S_{PO}^2, and S_E^2 are jointly independent.
2	$(p-1)S_P^2/\theta_P$ is a chi-squared random variable with $p-1$ degrees of freedom.
3	$(o-1)S_O^2/\theta_{PO}$ is a noncentral chi-squared random variable with $o-1$ degrees of freedom and noncentrality parameter $\lambda = \dfrac{pro\gamma_o}{2\theta_{PO}}$.
4	$(p-1)(o-1)S_{PO}^2/\theta_{PO}$ is a chi-squared random variable with $(p-1)(o-1)$ degrees of freedom.
5	$po(r-1)S_E^2/\theta_E$ is a chi-squared random variable with $po(r-1)$ degrees of freedom.
6	$\overline{Y}_{*j*}$ is a normal random variable with mean μ_j and variance $\dfrac{\theta_P + (o-1)\theta_{PO}}{por}$.
7	$\text{Cov}(\overline{Y}_{*j*}, \overline{Y}_{*j'*}) = \dfrac{\sigma_P^2}{p}$ for $j \neq j'$.
8	$\overline{Y}_{***}$ is a normal random variable with mean μ_Y and variance $\dfrac{\theta_P}{por}$.

Table 6.7. *Covariance structure for model* (6.3).

Condition	Covariance($Y_{ijk}, Y_{i'j'k'}$)
$i = i', j = j', k \neq k'$ (same part and same operator)	$\sigma_P^2 + \sigma_{PO}^2$
$i = i', j \neq j'$ (same part with different operators)	σ_P^2
$i \neq i', j = j'$ (same operator with different parts)	0
$i \neq i', j \neq j'$ (different parts and operators)	0

Table 6.8. *Gauge R&R parameters and point estimators for model* (6.3).

Gauge R&R notation	Model (6.3) representation	Point estimator
μ_Y	$\mu_Y = \dfrac{\Sigma_j \mu_j}{o}$	$\widehat{\mu}_Y = \overline{Y}_{***}$
μ_j	μ_j	$\widehat{\mu}_j = \overline{Y}_{*j*}$
γ_P	σ_P^2	$\widehat{\gamma}_P = \dfrac{S_P^2 - S_{PO}^2}{or}$
γ_M	$\gamma_o + \sigma_{PO}^2 + \sigma_E^2$	$\widehat{\gamma}_M = \{(o-1)S_O^2 + [o(p-1)+1]S_{PO}^2$ $+po(r-1)S_E^2\}/(por)$
γ_R	$\dfrac{\sigma_P^2}{\gamma_o + \sigma_{PO}^2 + \sigma_E^2}$	$\widehat{\gamma}_R = \dfrac{\widehat{\gamma}_P}{\widehat{\gamma}_M}$

Table 6.9. *Constants used in confidence intervals for model* (6.3). *Values are for* $\alpha = 0.05$, $p = 18$, $o = 3$, $r = 3$, *and* $n^* = 25$.

Constant	Definition	Value
G_1	$1 - F_{\alpha/2:\infty,p-1}$	0.4369
G_2	$1 - F_{\alpha/2:\infty,n^*}$	0.3849
G_3	$1 - F_{\alpha/2:\infty,(p-1)(o-1)}$	0.3457
G_4	$1 - F_{\alpha/2:\infty,po(r-1)}$	0.2211
H_1	$F_{1-\alpha/2:\infty,p-1} - 1$	1.247
H_2	$F_{1-\alpha/2:\infty,n^*} - 1$	0.9055
H_3	$F_{1-\alpha/2:\infty,(p-1)(o-1)} - 1$	0.7166
H_4	$F_{1-\alpha/2:\infty,po(r-1)} - 1$	0.3311
F_1	$F_{1-\alpha/2:p-1,(p-1)(o-1)}$	2.195
F_2	$F_{\alpha/2:p-1,(p-1)(o-1)}$	0.4042
F_3	$F_{1-\alpha/2:p-1,n^*}$	2.360
F_4	$F_{\alpha/2:p-1,n^*}$	0.3924
G_{13}	$\dfrac{(F_1-1)^2 - G_1^2F_1^2 - H_3^2}{F_1}$	−0.002058
H_{13}	$\dfrac{(1-F_2)^2 - H_1^2F_2^2 - G_3^2}{F_2}$	−0.04632

6.4.1 Interval for μ_Y

Results 1, 2, and 8 in Table 6.6 provide an exact confidence interval for μ_Y. This exact $100(1-\alpha)\%$ interval is

$$L = \overline{Y}_{***} - \sqrt{\frac{S_P^2 F_{1-\alpha:1,p-1}}{por}}$$

and

$$U = \overline{Y}_{***} + \sqrt{\frac{S_P^2 F_{1-\alpha:1,p-1}}{por}}. \tag{6.5}$$

6.4.2 Interval for μ_j

Based on Result 6 of Table 6.6, the estimator for μ_j is $\overline{Y}_{*j*}$ which has variance

$$\gamma_{\mu_j} = \frac{\theta_P + (o-1)\theta_{PO}}{por}.$$

A confidence interval for μ_j can be obtained using the Satterthwaite [57, 58] chi-squared approximation of the random variable $m\widehat{\gamma}_{\mu_j}/\gamma_{\mu_j}$, where

$$\widehat{\gamma}_{\mu_j} = \frac{S_P^2 + (o-1)S_{PO}^2}{por}.$$

The value of m obtained by matching the first two moments and replacing unknown parameters with their estimates is

$$m = \frac{[S_P^2 + (o-1)S_{PO}^2]^2}{\frac{S_P^4}{p-1} + \frac{(o-1)^2 S_{PO}^4}{(p-1)(o-1)}}. \tag{6.6}$$

An approximate $100(1-\alpha)\%$ interval for μ_j is then

$$L = \overline{Y}_{*j*} - \sqrt{\frac{F_{1-\alpha:1,m}[S_P^2 + (o-1)S_{PO}^2]}{por}}$$

and

$$U = \overline{Y}_{*j*} + \sqrt{\frac{F_{1-\alpha:1,m}[S_P^2 + (o-1)S_{PO}^2]}{por}}. \tag{6.7}$$

In determining $F_{1-\alpha:1,m}$ we truncate m in Equation (6.6) to the greatest integer less than or equal to m.

6.4.3 Interval for γ_P

The interval for γ_P shown in Equation (3.3) still holds under model (6.3) since it does not contain S_O^2. For convenience we repeat this interval below. The bounds of an approximate

$100(1-\alpha)\%$ confidence interval for γ_P are

$$L = \widehat{\gamma}_P - \frac{\sqrt{V_{LP}}}{or}$$

and

$$U = \widehat{\gamma}_P + \frac{\sqrt{V_{UP}}}{or}, \tag{6.8}$$

where

$$V_{LP} = G_1^2 S_P^4 + H_3^2 S_{PO}^4 + G_{13} S_P^2 S_{PO}^2,$$
$$V_{UP} = H_1^2 S_P^4 + G_3^2 S_{PO}^4 + H_{13} S_P^2 S_{PO}^2,$$

$\widehat{\gamma}_P$ is defined in Table 6.8, and G_1, G_3, H_1, H_3, G_{13}, and H_{13} are defined in Table 6.9. Negative bounds are increased to zero.

6.4.4 Interval for γ_M

The variability of the measurement process is $\gamma_M = \gamma_O + \sigma_{PO}^2 + \sigma_E^2$. Using the Graybill and Wang method [25] with n^* as the operator degrees of freedom, an approximate $100(1-\alpha)\%$ confidence interval for γ_M is

$$L = \widehat{\gamma}_M - \frac{\sqrt{V_{LM}}}{por}$$

and

$$U = \widehat{\gamma}_M + \frac{\sqrt{V_{UM}}}{por}, \tag{6.9}$$

where

$$V_{LM} = G_2^2(o-1)^2 S_O^4 + G_3^2[o(p-1)+1]^2 S_{PO}^4 + G_4^2 p^2 o^2 (r-1)^2 S_E^4,$$
$$V_{UM} = H_2^2(o-1)^2 S_O^4 + H_3^2[o(p-1)+1]^2 S_{PO}^4 + H_4^2 p^2 o^2 (r-1)^2 S_E^4,$$

$\widehat{\gamma}_M$ is defined in Table 6.8, and G_2, G_3, G_4, H_2, H_3, and H_4 are defined in Table 6.9.

6.4.5 Interval for γ_R

The bounds of the approximate $100(1-\alpha)\%$ confidence interval for γ_R based on results of Leiva and Graybill [42] are

$$L = \frac{p(1-G_1)(S_P^2 - F_1 S_{PO}^2)}{po(r-1)S_E^2 + (o-1)(1-G_1)F_3 S_O^2 + (po-o+1)S_{PO}^2}$$

and

$$U = \frac{p(1+H_1)(S_P^2 - F_2 S_{PO}^2)}{po(r-1)S_E^2 + (o-1)(1+H_1)F_4 S_O^2 + (po-o+1)S_{PO}^2}, \tag{6.10}$$

where G_1, H_1, F_1, F_2, F_3, and F_4 are defined in Table 6.9. Negative bounds are increased to zero.

Table 6.10. *GPQs for γ_O and gauge R&R parameters.*

Parameter	GPQ
γ_P	$\max\left[0, \dfrac{(p-1)s_P^2}{or W_1} - \dfrac{(p-1)(o-1)s_{PO}^2}{or W_3}\right]$
γ_O	$\sum_{t=1}^{o-1} \dfrac{Q_t}{ot(t+1)}$
γ_M	$\text{GPQ}(\gamma_O) + \dfrac{(p-1)(o-1)s_{PO}^2}{rW_3} + \dfrac{po(r-1)^2 s_E^2}{rW_4}$
γ_R	$\dfrac{\text{GPQ}(\gamma_P)}{\text{GPQ}(\gamma_M)}$

6.5 Generalized Intervals, Gauge R&R Parameters

Daniels, Burdick, and Quiroz [17] proposed a method for constructing a GPQ for γ_O and used it to construct GCIs for other parameters that contain γ_O. Let $\boldsymbol{Y}^*$ represent the $o \times 1$ vector of the $\overline{Y}_{*j*}$ and define $\boldsymbol{D} = \boldsymbol{\Delta}\boldsymbol{Y}^*$, where $\boldsymbol{\Delta}$ is an $(o-1) \times o$ matrix of the form

$$\Delta = \begin{pmatrix} 1 & -1 & 0 & \dots & 0 \\ 1 & 1 & -2 & \dots & 0 \\ \dots & \dots & \dots & \dots & \dots \\ 1 & 1 & 1 & \dots & (1-o) \end{pmatrix}. \tag{6.11}$$

Let D_t denote the tth element of $\boldsymbol{D}$. Table 6.10 reports GPQs for the gauge R&R parameters that contain γ_O, where

$$Q_t = d_t^2 - 2Z_1|R_t|\sqrt{\frac{t(t+1)(p-1)(o-1)s_{PO}^2}{prU_1}}, \tag{6.12}$$

$$R_t = d_t - Z_2\sqrt{\frac{t(t+1)(p-1)(o-1)s_{PO}^2}{prU_2}}, \tag{6.13}$$

and s_P^2, s_{PO}^2, s_E^2, and d_t represent the realized values of S_P^2, S_{PO}^2, S_E^2, and D_t. The random variables U_1 and U_2 are independent chi-squared random variables, each with $(p-1)(o-1)$ degrees of freedom, Z_1 and Z_2 are independent normal random variables with means zero and variances one, and W_1, W_3, and W_4 are independent chi-squared random variables with degrees of freedom $p-1$, $(p-1)(o-1)$, and $po(r-1)$, respectively. No GPQ is needed for μ_Y since Equation (6.5) provides an exact interval. Since γ_P does not contain γ_O, the GPQ for γ_P is that shown in Table 3.8.

The following process is used to compute generalized confidence intervals for the parameters shown in Table 6.10:

1. Compute $D_1, D_2, \ldots, D_{o-1}, S_P^2, S_{PO}^2$, and S_E^2 for the collected data and denote the realized values as $d_1, d_2, \ldots, d_{o-1}, s_P^2, s_{PO}^2$, and s_E^2, respectively.

2. Simulate the value of Q_t for $t = 1, 2, \ldots, o-1$ using Equations (6.12) and (6.13) by simulating independent values of U_1, U_2, Z_1, and Z_2.

3. Compute the GPQ defined in Table 6.10 for the parameter of interest. Use the results of step 2 and simulate independent values of W_1, W_3, and W_4.

4. Repeat steps 2 and 3 N times, where N is at least 100,000.

5. Order the N GPQ values created in step 4 from least to greatest.

6. Define the lower bound for a $100(1-\alpha)\%$ interval as the value in position $N \times (\alpha/2)$ of the ordered set in step 5. Define the upper bound as the value in position $N \times (1-\alpha/2)$ of this same ordered set.

Daniels, Burdick, and Quiroz [17] provide simulations that demonstrate that both the GCI and the MLS intervals given in Section 6.4 maintain the stated confidence level across a wide range of gauge R&R studies. The one exception is that confidence levels for lower bounds of γ_M in Equation (6.9) are sometimes lower than the stated level. With regard to interval length, the two sets of intervals are generally comparable. One exception is when the ratio $\gamma_O/\gamma_M \leq 0.2$. In this case, the generalized confidence intervals can be much shorter than the MLS intervals and are the recommended alternative.

6.6 Misclassification Rates

Since the population of measurements for each operator is a normal population, misclassification rates can be computed separately for each operator. Using model (6.1), the joint probability density function of Y_j and X is bivariate normal with mean vector $[\mu_j \quad \mu_P]'$ and covariance matrix

$$\begin{bmatrix} \gamma_P + \gamma_E & \gamma_P \\ \gamma_P & \gamma_P \end{bmatrix},$$

where for model (6.3), $\gamma_P = \sigma_P^2$ and $\gamma_E = \sigma_{PO}^2 + \sigma_E^2$. Using this model, we construct confidence intervals for the misclassification rates using the GPQs shown in Table 6.11, where $\overline{y}_{***}, \overline{y}_{*j*}, s_P^2, s_{PO}^2$, and s_E^2 represent the realized values of $\overline{Y}_{***}, \overline{Y}_{*j*}, S_P^2, S_{PO}^2$, and S_E^2. The random variables W_1, W_3, and W_4 are independent chi-squared random variables with degrees of freedom $p-1$, $(p-1)(o-1)$, and $po(r-1)$, respectively, and Z is a normal random variable with mean zero and variance one.

The reader is reminded that $\overline{Y}_{***}$ estimates μ_P because we assume $\mu_Y = \mu_P$. If this assumption is not valid, then an independent estimate of μ_P is needed to estimate misclassification rates.

Table 6.11. *GPQs for misclassification rates in model* (6.3).

Parameter	GPQ
μ_j	$\overline{y}_{*j*} - Z\sqrt{\frac{(p-1)s_P^2}{por\,W_1} + \frac{(p-1)(o-1)^2 s_{PO}^2}{por\,W_3}}$
μ_P	$\overline{y}_{***} - Z\sqrt{\frac{(p-1)s_P^2}{por\,W_1}}$
$\gamma_P + \gamma_E$	$\frac{(p-1)s_P^2}{or\,W_1} + \frac{(p-1)(o-1)^2 s_{PO}^2}{or\,W_3} + \frac{po(r-1)^2 s_E^2}{r\,W_4}$
γ_P	$\max\left[\epsilon, \frac{(p-1)s_P^2}{or\,W_1} - \frac{(p-1)(o-1)s_{PO}^2}{or\,W_3}\right]$

Table 6.12. *ANOVA for two-factor mixed model example.*

Source of variation	Degrees of freedom	Mean square
Parts (P)	17	2.056
Operators (O)	2	2.505
P×O	34	0.09395
Replicates	108	0.07478

6.7 Numerical Example

We demonstrate the formulas in this section by performing an analysis of the full data set represented in Table 6.1. The ANOVA is shown in Table 6.12. Remember that operator is a fixed effect in this model. The mean of all the observations is $\overline{y}_{***} = -40.146$, and the means for each operator are $\overline{y}_{*1*} = -39.972$, $\overline{y}_{*2*} = -40.387$, and $\overline{y}_{*3*} = -40.079$. The specification limits are $LSL = -41$ and $USL = -33$.

We begin by computing n^*. Using Equation (6.4),

$$\widehat{\lambda} = \frac{o-1}{2}\left[\frac{S_O^2}{S_{PO}^2}\left\{\frac{(p-1)(o-1)-2}{(p-1)(o-1)}\right\} - 1\right]$$

$$= \frac{2}{2}\left[\frac{2.505}{0.09395}\left\{\frac{34-2}{34}\right\} - 1\right]$$

$$= 24.09$$

and

$$n^* = \frac{[(o-1)+2\widehat{\lambda}]^2}{(o-1)+4\widehat{\lambda}}$$
$$= \frac{[2+2(24.09)]^2}{2+4(24.09)}$$
$$= 25.6$$
$$= 25 \text{ (truncated).}$$

To construct two-sided 95% confidence intervals, we use the constants shown in Table 6.9 with $\alpha = 0.05$, $p = 18$ parts, $o = 3$ operators, $r = 3$ replicates, and $n^* = 25$.

6.7.1 Interval for μ_Y

The point estimate for μ_Y is $\overline{y}_{***} = -40.146$. The confidence interval is shown in Equation (6.5). With $\alpha = 0.05$, $F_{1-\alpha:1,p-1} = F_{0.95:1,17} = 4.451$. Hence the 95% confidence interval for μ_Y is

$$L = \overline{Y}_{***} - \sqrt{\frac{S_P^2 F_{1-\alpha:1,p-1}}{por}}$$
$$= -40.146 - \sqrt{\frac{2.056(4.451)}{18(3)(3)}}$$
$$= -40.4$$

and

$$U = \overline{Y}_{***} + \sqrt{\frac{S_P^2 F_{1-\alpha:1,p-1}}{por}}$$
$$= -40.146 + \sqrt{\frac{2.056(4.451)}{18(3)(3)}}$$
$$= -39.9.$$

6.7.2 Interval for μ_j

The confidence interval for μ_j is obtained by first computing m from Equation (6.6) as

$$m = \frac{[S_P^2 + (o-1)S_{PO}^2]^2}{\frac{S_P^4}{p-1} + \frac{(o-1)^2 S_{PO}^4}{(p-1)(o-1)}}$$
$$= \frac{[2.056 + 2(0.09395)]^2}{\frac{(2.056)^2}{17} + \frac{(2)^2(0.09395)^2}{34}}$$
$$= 20.17$$
$$= 20 \text{ (truncated).}$$

Table 6.13. *95% confidence intervals for operator effects.*

Parameter	Estimate	Lower bound	Upper bound
μ_1	−39.972	−40.22	−39.73
μ_2	−40.387	−40.63	−40.14
μ_3	−40.079	−40.32	−39.83

We now compute the 95% confidence interval for μ_1 using Equation (6.7) with $F_{1-\alpha:1,m} = F_{0.95:1,20} = 4.3512$ to be

$$\begin{aligned} L &= \overline{Y}_{*j*} - \sqrt{\frac{F_{1-\alpha:1,m}[S_P^2 + (o-1)S_{PO}^2]}{por}} \\ &= -39.972 - \sqrt{\frac{4.3512[2.056 + 2(0.09395)]}{18(3)(3)}} \\ &= -40.22 \end{aligned}$$

and

$$\begin{aligned} U &= \overline{Y}_{*j*} + \sqrt{\frac{F_{1-\alpha:1,m}[S_P^2 + (o-1)S_{PO}^2]}{por}} \\ &= -39.972 + \sqrt{\frac{4.3512[2.056 + 2(0.09395)]}{18(3)(3)}} \\ &= -39.73. \end{aligned}$$

The 95% confidence intervals for μ_2 and μ_3 obtained in the same manner are given in Table 6.13. We note that operator 2 has a somewhat lesser average than the other two operators. In Section 6.8 we show how to determine if the true operator means differ.

6.7.3 Interval for γ_P

The point estimate for γ_P is

$$\begin{aligned} \widehat{\gamma}_P &= \frac{S_P^2 - S_{PO}^2}{or} \\ &= \frac{2.056 - 0.09395}{9} \\ &= 0.218. \end{aligned}$$

The lower and upper confidence bounds are shown in Equation (6.8). In this example,

$$\begin{aligned} V_{LP} &= G_1^2 S_P^4 + H_3^2 S_{PO}^4 + G_{13} S_P^2 S_{PO}^2 \\ &= (0.4369)^2(2.056)^2 + (0.7166)^2(0.09395)^2 \\ &\quad + (-0.002058)(2.056)(0.09395) \\ &= 0.811 \end{aligned}$$

and

$$\begin{aligned} V_{UP} &= H_1^2 S_P^4 + G_3^2 S_{PO}^4 + H_{13} S_P^2 S_{PO}^2 \\ &= (1.247)^2(2.056)^2 + (0.3457)^2(0.09395)^2 \\ &\quad + (-0.04632)(2.056)(0.09395) \\ &= 6.57. \end{aligned}$$

Substituting this information into Equation (6.8) yields the 95% confidence interval for γ_P,

$$L = 0.218 - \frac{\sqrt{0.811}}{9} = 0.118$$

and

$$U = 0.218 + \frac{\sqrt{6.57}}{9} = 0.503. \tag{6.14}$$

6.7.4 Interval for γ_M

The point estimate for γ_M is

$$\begin{aligned} \widehat{\gamma}_M &= \frac{(o-1)S_O^2 + [o(p-1)+1]S_{PO}^2 + po(r-1)S_E^2}{por} \\ &= \frac{2(2.505) + 52(0.09395) + 108(0.07478)}{162} \\ &= 0.1109. \end{aligned}$$

The lower and upper confidence bounds for γ_M are shown in Equation (6.9). In this example,

$$\begin{aligned} V_{LM} &= G_2^2(o-1)^2 S_O^4 + G_3^2[o(p-1)+1]^2 S_{PO}^4 + G_4^2 p^2 o^2 (r-1)^2 S_E^4 \\ &= (0.3849)^2(2)^2(2.505)^2 + (0.3457)^2(52)^2(0.09395)^2 \\ &\quad + (0.2211)^2(108)^2(0.07478)^2 \\ &= 9.760 \end{aligned}$$

and

$$\begin{aligned} V_{UM} &= H_2^2(o-1)^2 S_O^4 + H_3^2[o(p-1)+1]^2 S_{PO}^4 + H_4^2 p^2 o^2 (r-1)^2 S_E^4 \\ &= (0.9055)^2(2)^2(2.505)^2 + (0.7166)^2(52)^2(0.09395)^2 \\ &\quad + (0.3311)^2(108)^2(0.07478)^2 \\ &= 39.99. \end{aligned}$$

Substituting this information into Equation (6.9) yields the 95% confidence interval for γ_M,

$$L = 0.1109 - \frac{\sqrt{9.760}}{162} = 0.092$$

and

$$U = 0.1109 + \frac{\sqrt{39.99}}{162} = 0.150. \tag{6.15}$$

To compute the generalized confidence interval for γ_M using Table 6.10, we note $d_1 = \overline{y}_{*1*} - \overline{y}_{*2*} = -39.972 - (-40.387) = 0.415$ and $d_2 = \overline{y}_{*1*} + \overline{y}_{*2*} - 2\overline{y}_{*3*} = -39.972 - 40.387 - 2(-40.079) = -0.201$. Our computed 95% GCI is from $L = 0.086$ to $U = 0.147$.

6.7.5 Interval for γ_R

The point estimate for γ_R is

$$\begin{aligned}\widehat{\gamma}_R &= \frac{\widehat{\gamma}_P}{\widehat{\gamma}_M} \\ &= \frac{0.2180}{0.1109} \\ &= 1.966.\end{aligned}$$

The $100(1-\alpha)\%$ confidence interval for γ_R using Equation (6.10) is

$$\begin{aligned}L &= \frac{p(1-G_1)(S_P^2 - F_1 S_{PO}^2)}{po(r-1)S_E^2 + (o-1)(1-G_1)F_3 S_O^2 + (po-o+1)S_{PO}^2} \\ &= \frac{18(0.5631)(2.056 - 2.195(0.09395))}{108(0.07478) + 2(0.5631)(2.360)(2.505) + 52(0.09395)} \\ &= 0.956\end{aligned}$$

and

$$\begin{aligned}U &= \frac{p(1+H_1)(S_P^2 - F_2 S_{PO}^2)}{po(r-1)S_E^2 + (o-1)(1+H_1)F_4 S_O^2 + (po-o+1)S_{PO}^2} \\ &= \frac{18(2.247)(2.056 - 0.4042(0.09395))}{108(0.07478) + 2(2.247)(0.3924)(2.505) + 52(0.09395))} \\ &= 4.70.\end{aligned} \tag{6.16}$$

The 95% GCI computed with the GPQ shown in Table 6.10 is from $L = 0.969$ to $U = 4.59$.

6.7.6 Interval for PTR

Using Equation (1.2) with $k = 6$, a point estimate for PTR is

$$\frac{k\sqrt{\widehat{\gamma}_M}}{USL - LSL} = \frac{6\sqrt{0.1109}}{(-33) - (-41)} = 0.250.$$

A 95% confidence interval for PTR based on the bounds calculated for γ_M in Equation (6.15) is

$$L = \frac{6\sqrt{0.092}}{(-33) - (-41)} = 0.227$$

and

$$U = \frac{6\sqrt{0.150}}{(-33) - (-41)} = 0.290.$$

6.7.7 Interval for SNR

Based on Equation (1.3), the point estimate for SNR is

$$\sqrt{2\widehat{\gamma}_R} = \sqrt{2(1.966)} = 1.98.$$

A 95% confidence interval for the SNR based on the bounds for γ_R calculated in Equation (6.16) is

$$L = \sqrt{2(0.956)} = 1.38$$

and

$$U = \sqrt{2(4.70)} = 3.07.$$

6.7.8 Interval for C_p

The 95% confidence interval for C_p using the computed interval for γ_P in Equation (6.14) is

$$L = \frac{(-33) - (-41)}{6\sqrt{0.503}} = 1.88$$

and

$$U = \frac{(-33) - (-41)}{6\sqrt{0.118}} = 3.88.$$

6.7.9 Intervals for Misclassification Rates

Table 6.14 gives 95% confidence intervals for the misclassification rates of each operator. These intervals were computed using the algorithm described in Section 3.4 with the GPQ values defined in Table 6.11 and $\epsilon = 0.001$. Each interval is based on 100,000 GPQ values. Overall misclassification rates can be obtained by averaging bounds across the three operators.

Based on the misclassification indexes, there is no evidence that any of the operators are doing much better than a chance process. Operator 2, in particular, has a very high false failure rate. This suggests some effort is needed to better calibrate the three operators.

Table 6.14. *95% intervals for misclassification rates. (Bounds for δ, δ_c, β, and β_c have been multiplied by* 10^6.*)*

Misclassification rate	Operator 1	Operator 2	Operator 3
δ	$L = 4{,}736$	$L = 49{,}221$	$L = 9{,}665$
	$U = 26{,}759$	$U = 158{,}866$	$U = 46{,}476$
δ_c	$L = 4{,}764$	$L = 49{,}621$	$L = 9{,}728$
	$U = 30{,}222$	$U = 178{,}278$	$U = 52{,}520$
δ_{index}	$L = 0.148$	$L = 0.962$	$L = 0.272$
	$U = 1.45$	$U = 13.7$	$U = 2.77$
β	$L = 2{,}550$	$L = 565$	$L = 1{,}970$
	$U = 55{,}194$	$U = 9{,}733$	$U = 39{,}416$
β_c	$L = 353{,}960$	$L = 55{,}933$	$L = 252{,}767$
	$U = 597{,}735$	$U = 141{,}789$	$U = 459{,}193$
β_{index}	$L = 0.411$	$L = 0.064$	$L = 0.293$
	$U = 0.601$	$U = 0.143$	$U = 0.461$

Table 6.15. *Summary of MLS and generalized confidence intervals.*

Parameter	MLS	GCI
γ_P	$L = 0.118$	$L = 0.117$
	$U = 0.503$	$U = 0.501$
γ_M	$L = 0.092$	$L = 0.086$
	$U = 0.150$	$U = 0.147$
γ_R	$L = 0.956$	$L = 0.969$
	$U = 4.70$	$U = 4.59$
PTR	$L = 0.227$	$L = 0.220$
	$U = 0.290$	$U = 0.288$
SNR	$L = 1.38$	$L = 1.39$
	$U = 3.07$	$U = 3.03$
C_p	$L = 1.88$	$L = 1.88$
	$U = 3.88$	$U = 3.90$

6.7.10 Conclusions

Table 6.15 gives computed intervals used to determine the capability of the measurement system. Figure 6.1 presents the R&R graph. It appears the measurement system satisfies the PTR criterion, but all the values in the SNR confidence interval are less than the desired value of five. This likely occurs because the process has a very high capability measure ($C_p \geq 1.88$). The plots shown in Figure 6.2 suggest a slight skewness in the residuals and some increase in spread for the latter observations (parts 15–18). Figure 6.3 suggests

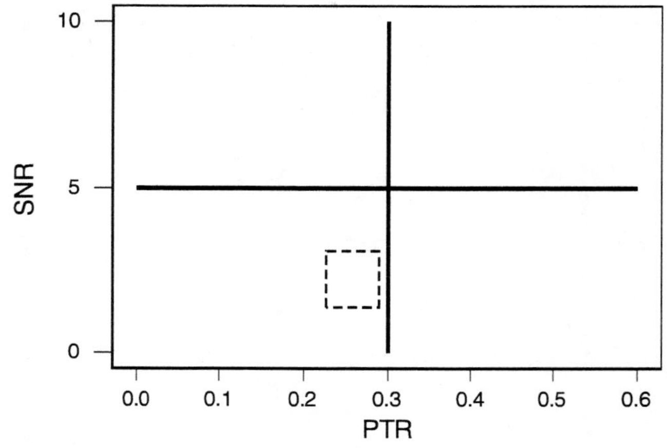

Figure 6.1. *R&R graph for mixed example.*

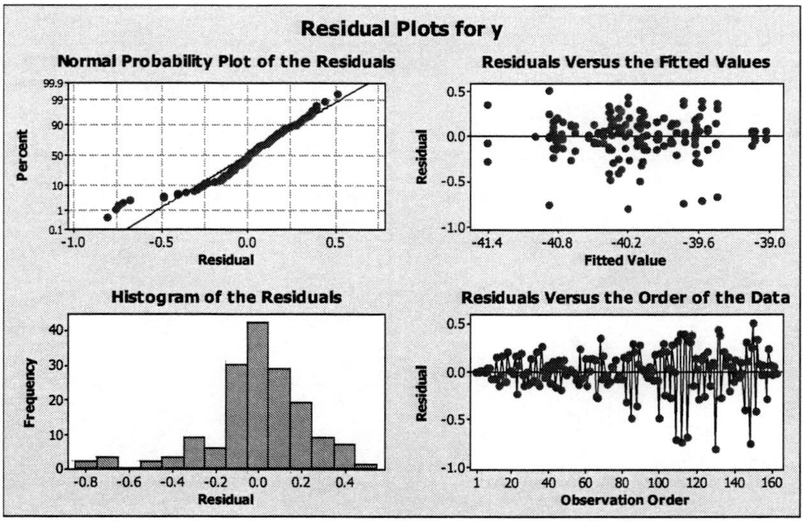

Figure 6.2. *Residual plots for the two-factor mixed model.*

that measurements from Operator 2 have slightly greater variability than for the other two operators. Additionally, as shown in Figures 6.4 and 6.5, the variability is increasing over time (assuming parts are ordered sequentially). It is of interest to note in Figure 6.5 that both of the values outside the control limits in the range chart are for Operator 2. In the $\overline{X}$ chart the majority of the sample means are outside of the control limits. This provides evidence that the system has some ability to classify parts and is consistent with the PTR criterion. As noted earlier, the misclassification rates suggest the system could be improved by adjustment of the operators (testers).

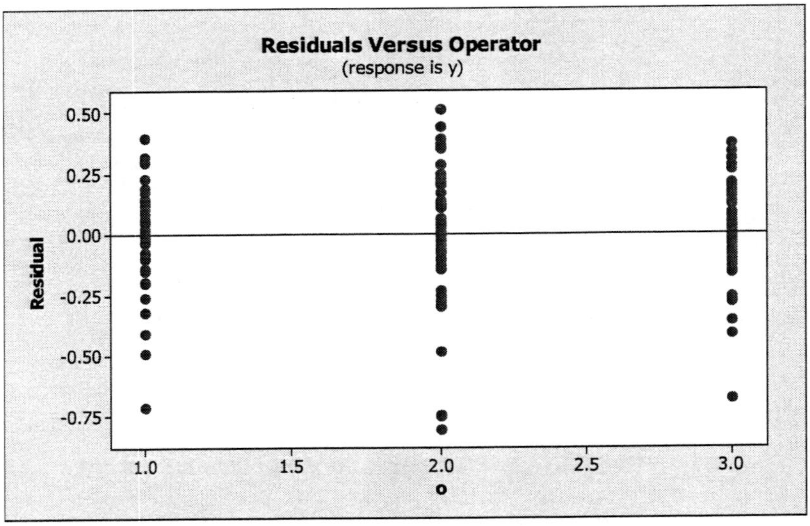

Figure 6.3. *Residual versus operator for two-factor mixed model.*

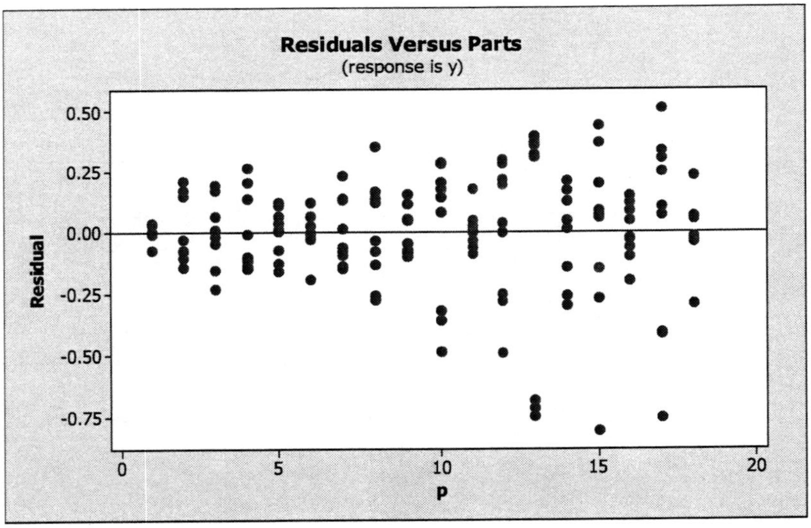

Figure 6.4. *Residual versus parts for two-factor mixed model.*

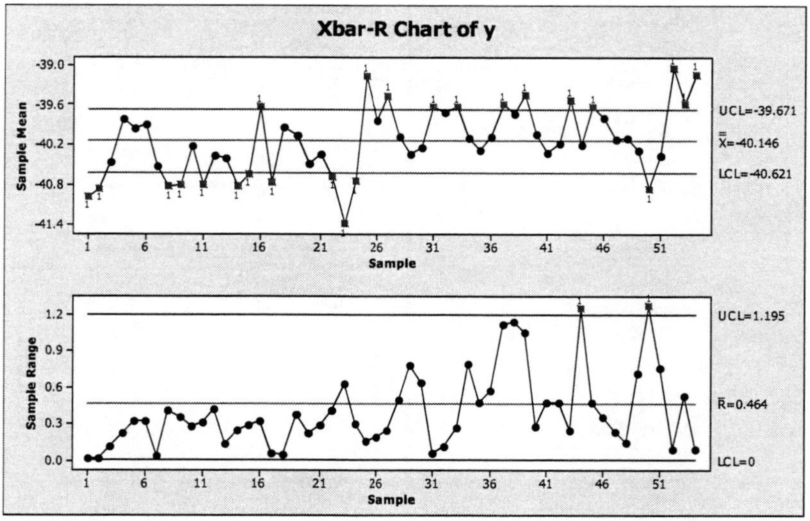

Figure 6.5. $\overline{X}$ *and R control charts for the two-factor mixed model.*

6.8 Confidence Intervals for Contrasts of μ_j

When using a mixed model, one will likely want to compare the fixed (operator) effects. This is particularly useful in a gauge R&R study because it might suggest ways to reduce variability among operators. The easiest way to examine operator effects is to construct confidence intervals on the $o(o-1)/2$ pairwise comparisons of the μ_j. For example, if we have $o = 3$ operators, we compute confidence intervals for the $o(o-1)/2 = 3$ comparisons $\mu_1 - \mu_2$, $\mu_1 - \mu_3$, and $\mu_2 - \mu_3$.

A pairwise comparison can be represented more generally as the linear contrast $\omega = \Sigma_j c_j \mu_j$, where $\Sigma_j c_j = 0$. For example, the pairwise comparison $\mu_1 - \mu_2$ is written in this form with $c_1 = 1$, $c_2 = -1$, and $c_3 = 0$. The best linear unbiased estimator for ω is $\widehat{\omega} = \Sigma_j c_j \overline{Y}_{*j*}$. Based on Results 6 and 7 of Table 6.6, the variance of $\widehat{\omega}$ is

$$\frac{(\Sigma_j c_j^2)\theta_{PO}}{pr}.$$

Thus, an exact $100(1-\alpha)\%$ interval for ω is

$$L = \widehat{\omega} - \sqrt{\frac{(\Sigma_j c_j^2) S_{PO}^2 F_{1-\alpha:1,(p-1)(o-1)}}{pr}}$$

and

$$U = \widehat{\omega} + \sqrt{\frac{(\Sigma_j c_j^2) S_{PO}^2 F_{1-\alpha:1,(p-1)(o-1)}}{pr}}. \tag{6.17}$$

Note that $\Sigma_j c_j^2 = 2$ for any pairwise comparison.

Table 6.16. *95% confidence intervals for operator contrasts.*

Contrast	Estimate	Lower bound	Upper bound
$\mu_1 - \mu_2$	0.415	0.30	0.53
$\mu_1 - \mu_3$	0.107	−0.013	0.23
$\mu_2 - \mu_3$	−0.308	−0.43	−0.19

To demonstrate, we compare the $o = 3$ operators in the numerical example of the previous section. Here $\overline{y}_{*1*} = -39.972$, $\overline{y}_{*2*} = -40.387$, $\overline{y}_{*3*} = -40.079$, $p = 18$, $r = 3$, and $s^2_{PO} = 0.09395$. Consider the contrast $\omega = \mu_1 - \mu_2$ so that $c_1 = 1$, $c_2 = -1$, and $c_3 = 0$. The computed value for $\widehat{\omega} = -39.972 - (-40.387) = 0.415$. If we desire a 95% confidence interval, then $\alpha = 0.05$ and $F_{1-\alpha:1,(p-1)(o-1)} = F_{0.95:1,34} = 4.1300$. The computed 95% confidence interval for ω using Equation (6.17) is

$$\begin{aligned} L &= \widehat{\omega} - \sqrt{\frac{(\Sigma_j c_j^2) S^2_{PO} F_{1-\alpha:1,(p-1)(o-1)}}{pr}} \\ &= 0.415 - \sqrt{\frac{2(0.09395)(4.1300)}{18(3)}} \\ &= 0.30 \end{aligned}$$

and

$$\begin{aligned} U &= \widehat{\omega} + \sqrt{\frac{(\Sigma_j c_j^2) S^2_{PO} F_{1-\alpha:1,(p-1)(o-1)}}{pr}} \\ &= 0.415 + \sqrt{\frac{2(0.09395)(4.1300)}{18(3)}} \\ &= 0.53. \end{aligned}$$

Table 6.16 reports all three pairwise comparisons for the example. Since all values in the interval for $\mu_1 - \mu_2$ are positive and all values in the interval for $\mu_2 - \mu_3$ are negative, these results suggest Operator 2 has a significantly lesser average than the other two operators. There is no evidence that μ_1 and μ_3 differ since the confidence interval for $\mu_1 - \mu_3$ contains both negative and positive values. Given that Operator 2 also has the greatest false failure rate (see Table 6.14), an adjustment that increases μ_2 will likely improve the measurement system.

6.9 The Two-Factor Mixed Model with No Interaction

We now provide brief results for the two-factor mixed model with no interaction. The two-factor random model with no interaction was the topic of Chapter 5. The two-factor mixed

Table 6.17. *ANOVA for model* (6.18).

Source of variation	Degrees of freedom	Mean square	Expected mean square
Parts (P)	$p-1$	S_P^2	$\theta_P = \sigma_E^2 + or\sigma_P^2$
Operators (O)	$o-1$	S_O^2	$\theta_O = \sigma_E^2 + pr\left(\frac{o\gamma_o}{o-1}\right)$
Replicates (E)	$por-p-o+1$	S_E^2	$\theta_E = \sigma_E^2$

Table 6.18. *Distributional results for model* (6.18).

Result	
1	$\overline{Y}_{*j*}$, S_P^2, S_O^2, and S_E^2 are jointly independent.
2	$(p-1)S_P^2/\theta_P$ is a chi-squared random variable with $p-1$ degrees of freedom.
3	$(o-1)S_O^2/\theta_E$ is a noncentral chi-squared random variable with $o-1$ degrees of freedom and noncentrality parameter $\lambda = \dfrac{pro\gamma_o}{2\theta_E}$.
4	$(por-p-o+1)S_E^2/\theta_E$ is a chi-squared random variable with $(por-p-o+1)$ degrees of freedom.
5	$\overline{Y}_{*j*}$ is a normal random variable with mean μ_j and variance $\dfrac{\theta_P+(o-1)\theta_E}{por}$.
6	$\mathrm{Cov}(\overline{Y}_{*j*}, \overline{Y}_{*j'*}) = \dfrac{\sigma_P^2}{p}$ for $j \neq j'$.
7	$\overline{Y}_{***}$ is a normal random variable with mean μ_Y and variance $\dfrac{\theta_P}{por}$.

model with no interactionis

$$Y_{ijk} = P_i + \mu_j + E_{ijk}, \qquad (6.18)$$
$$i = 1,\ldots,p, \quad j = 1,\ldots,o, \quad k = 1,\ldots,r,$$

where the μ_j are constants specific to a particular operator and P_i and E_{ijk} are jointly independent normal random variables with means of zero and variances σ_P^2 and σ_E^2, respectively.

The ANOVA for model (6.18) is shown in Table 6.17, where $\gamma_o = \Sigma_j(\mu_j - \mu_Y)^2/o$ and $\mu_Y = \Sigma_j\mu_j/o$. The definitions for the mean squares and means are the same as those shown in Table 5.2. Table 6.18 gives distributional properties based on the assumptions of model (6.18).

6.9.1 Intervals for R&R Parameters

The confidence interval for γ_P is still Equation (5.3) since it does not involve γ_O. One approach for constructing confidence intervals for functions of γ_O is to use the formulas in Chapter 5, replacing the operator degrees of freedom with n^*, where

$$n^* = \frac{[(o-1)+2\widehat{\lambda}]^2}{(o-1)+4\widehat{\lambda}} \tag{6.19}$$

and

$$\widehat{\lambda} = \frac{o-1}{2}\left[\frac{S_O^2}{S_E^2}\left\{\frac{(por-p-o+1)-2}{(por-p-o+1)}\right\} - 1\right]. \tag{6.20}$$

The confidence interval for $\gamma_M = \gamma_O + \sigma_E^2$ is then

$$L = \widehat{\gamma}_M - \frac{\sqrt{V_{LM}}}{por}$$

and

$$U = \widehat{\gamma}_M + \frac{\sqrt{V_{UM}}}{por},$$

where

$$\widehat{\gamma}_M = \frac{(o-1)S_O^2 + (por-o+1)S_E^2}{por},$$

$$V_{LM} = G_2^2(o-1)^2 S_O^4 + G_3^2(por-o+1)^2 S_E^4,$$

$$V_{UM} = H_2^2(o-1)^2 S_O^4 + H_3^2(por-o+1)^2 S_E^4,$$

and G_2, G_3, H_2, and H_3 are defined in Table 5.5 with n^* defined in Equation (6.19) replacing $o-1$ in G_2 and H_2. Finally, the confidence interval for γ_R is

$$L = \frac{p(1-G_1)S_P^4 - pS_P^2S_E^2 + p[F_1 - (1-G_1)F_1^2]S_E^4}{(por-o+1)S_P^2S_E^2 + (o-1)(1-G_1)F_3S_P^2S_O^2}$$

and

$$U = \frac{p(1+H_1)S_P^4 - pS_P^2S_E^2 + p[F_2 - (1+H_1)F_2^2]S_E^4}{(por-o+1)S_P^2S_E^2 + (o-1)(1+H_1)F_4S_P^2S_O^2},$$

where G_1, H_1, F_1, F_2, F_3, and F_4 are defined in Table 5.5 with n^* defined in Equation (6.19) replacing $o-1$ in F_3 and F_4. Negative bounds are increased to zero.

Table 6.19. *GPQs for model* (6.18).

Parameter	GPQ
γ_P	$\max\left[0, \frac{(p-1)s_P^2}{or W_1} - \frac{(por-p-o+1)s_E^2}{or W_3}\right]$
γ_O	$\sum_{t=1}^{o-1} \frac{Q_t}{ot(t+1)}$
γ_M	$\text{GPQ}(\gamma_O) + \frac{(por-p-o+1)s_E^2}{W_3}$
γ_R	$\frac{\text{GPQ}(\gamma_P)}{\text{GPQ}(\gamma_M)}$

Alternatively, one can construct generalized confidence intervals using the same approach presented in Section 6.5. Here we define $\boldsymbol{Y}^*$ as the $o \times 1$ vector of the $\overline{Y}_{*j*}$ and define $\boldsymbol{D} = \boldsymbol{\Delta}\boldsymbol{Y}^*$, where $\boldsymbol{\Delta}$ is defined in Equation (6.11). Let D_t denote the tth element of $\boldsymbol{D}$. Table 6.19 gives GPQs, where

$$Q_t = d_t^2 - 2Z_1|R_t|\sqrt{\frac{t(t+1)(por-p-o+1)s_E^2}{prU_1}},$$

$$R_t = d_t - Z_2\sqrt{\frac{t(t+1)(por-p-o+1)s_E^2}{prU_2}},$$

and s_P^2, s_E^2, and d_t represent the realized values of S_P^2, S_E^2, and D_t. The random variables U_1 and U_2 are independent chi-squared random variables, each with $(por - p - o + 1)$ degrees of freedom, Z_1 and Z_2 are independent normal random variables with means zero and variances one, and W_1 and W_3 are independent chi-squared random variables with degrees of freedom $p - 1$ and $(por - p - o + 1)$, respectively.

6.9.2 Intervals for Misclassification Rates

When operator is a fixed effect, it is necessary to compute misclassification rates individually for each of the o operators. These misclassification rates can be computed using the GPQs shown in Table 6.20, where $\overline{y}_{***}$, $\overline{y}_{*j*}$, s_P^2, and s_E^2 represent the realized values of $\overline{Y}_{***}$, $\overline{Y}_{*j*}$, S_P^2, and S_E^2. (Recall we are assuming $\mu_Y = \mu_P$.) The random variables W_1 and W_3 are independent chi-squared random variables with degrees of freedom $p - 1$ and $(por - p - o + 1)$, respectively, and Z is a normal random variable with mean zero and variance one.

Table 6.20. *GPQs for misclassification rates in model* (6.18).

Parameter	GPQ
μ_j	$\overline{y}_{*j*} - Z\sqrt{\frac{(p-1)s_P^2}{por\,W_1} + \frac{(o-1)(por-p-o+1)s_E^2}{por\,W_3}}$
μ_P	$\overline{y}_{***} - Z\sqrt{\frac{(p-1)s_P^2}{por\,W_1}}$
$\gamma_P + \gamma_E = \sigma_P^2 + \sigma_E^2$	$\frac{(p-1)s_P^2}{or\,W_1} + \frac{(or-1)(por-p-o+1)s_E^2}{or\,W_3}$
γ_P	$\max\left[\epsilon, \frac{(p-1)s_P^2}{or\,W_1} - \frac{(por-p-o+1)s_E^2}{or\,W_3}\right]$

Table 6.21. *Formulas for confidence intervals on functions of* μ_j *in model* (6.18).

Parameter	Equation	Modification
μ_Y	(6.5)	None
μ_j	(6.7)	Replace S_{PO}^2 with S_E^2 Replace $(p-1)(o-1)$ with $(por-p-o+1)$ in Equation (6.6)
$\omega = \Sigma_j c_j \mu_j$	(6.17)	Replace S_{PO}^2 with S_E^2 Replace $F_{1-\alpha:1,(p-1)(o-1)}$ with $F_{1-\alpha:1,por-p-o+1}$

6.9.3 Intervals for μ_Y, μ_j, and $\omega = \Sigma_j c_j \mu_j$

The confidence intervals for μ_Y, μ_j, and $\omega = \Sigma_j c_j \mu_j$ are the same ones used for model (6.3) with the exception that S_{PO}^2 is replaced with S_E^2, and the degrees of freedom $(p-1)(o-1)$ is replaced with $por - p - o + 1$. These modifications are summarized in Table 6.21.

6.9.4 Numerical Example

We complete this section by computing confidence intervals for the data shown in Table 6.1 under model (6.18). The ANOVA for the model with interaction is shown in Table 6.22. As stated in Section 3.8.1, the null hypothesis $H_0 : \sigma_{PO}^2 = 0$ is rejected in favor of $H_a : \sigma_{PO}^2 > 0$ if $S_{PO}^2/S_E^2 > F_{1-\alpha/2:(p-1)(o-1),po(r-1)}$. This test for the data in Table 6.22 is $s_{PO}^2/s_E^2 = 0.09395/0.07478 = 1.26$. The critical value is $F_{1-\alpha/2:(p-1)(o-1),po(r-1)} = F_{0.975:34,108} = 1.67$ and so the null hypothesis cannot be rejected. The p-value for the test is 0.1891. This

Table 6.22. *ANOVA for two-factor mixed model with interaction.*

Source of variation	Degrees of freedom	Mean square
Parts (P)	17	2.056
Operators (O)	2	2.505
P×O	34	0.09395
Replicates	108	0.07478

Table 6.23. *ANOVA for two-factor mixed model with no interaction.*

Source of variation	Degrees of freedom	Mean square
Parts	17	2.056
Operators	2	2.505
Replicates	142	0.07937

test suggests that the interaction term is not needed and that model (6.18) is more appropriate than the interaction model for this data set.

The ANOVA for model (6.18) is shown in Table 6.23. From Equation (6.20),

$$\begin{aligned}\widehat{\lambda} &= \frac{o-1}{2}\left[\frac{S_O^2}{S_E^2}\left\{\frac{(por-p-o+1)-2}{(por-p-o+1)}\right\}-1\right]\\ &= \frac{3-1}{2}\left[\frac{2.505}{0.07937}\left\{\frac{142-2}{142}\right\}-1\right]\\ &= 30.12.\end{aligned}$$

Thus, from Equation (6.19),

$$\begin{aligned}n^* &= \frac{[(o-1)+2\widehat{\lambda}]^2}{(o-1)+4\widehat{\lambda}}\\ &= \frac{[(3-1)+2(30.12)]^2}{(3-1)+4(30.12)}\\ &= 31.6\\ &= 31 \text{ (truncated).}\end{aligned}$$

The computed confidence intervals are reported in Table 6.24. The generalized confidence intervals are based on 100,000 GPQ values with $\epsilon = 0.001$. The confidence intervals for the misclassification rates are shown in Table 6.25, and the intervals for functions of the μ_j are shown in Table 6.26. These computations use $\alpha = 0.05$, $n^* = 31$, $p = 18$, $o = 3$, $r = 3$,

Table 6.24. *Summary of MLS and generalized confidence intervals. (See Section 1.8 for a description of computer programs to perform these computations.)*

Parameter	MLS	GCI
γ_P	$L = 0.120$	$L = 0.120$
	$U = 0.505$	$U = 0.501$
γ_M	$L = 0.090$	$L = 0.085$
	$U = 0.142$	$U = 0.139$
γ_R	$L = 1.02$	$L = 1.03$
	$U = 4.75$	$U = 4.69$
PTR	$L = 0.225$	$L = 0.219$
	$U = 0.282$	$U = 0.280$
SNR	$L = 1.43$	$L = 1.44$
	$U = 3.08$	$U = 3.06$
C_p	$L = 1.88$	$L = 1.88$
	$U = 3.85$	$U = 3.85$

Table 6.25. 95% *intervals for misclassification rates. (Bounds for δ, δ_c, β, and β_c have been multiplied by* 10^6.*)*

Misclassification rate	Operator 1	Operator 2	Operator 3
δ	$L = 4{,}540$	$L = 48{,}952$	$L = 9{,}430$
	$U = 24{,}914$	$U = 155{,}759$	$U = 44{,}427$
δ_c	$L = 4{,}568$	$L = 49{,}340$	$L = 9{,}500$
	$U = 28{,}328$	$U = 175{,}759$	U = 50,422
δ_{index}	$L = 0.141$	$L = 0.958$	$L = 0.263$
	$U = 1.28$	$U = 12.8$	$U = 2.47$
β	$L = 2{,}675$	$L = 564$	$L = 2{,}123$
	$U = 55{,}471$	$U = 9{,}394$	$U = 39{,}588$
β_c	$L = 352{,}356$	$L = 54{,}078$	$L = 249{,}998$
	$U = 597{,}086$	$U = 133{,}670$	$U = 454{,}437$
β_{index}	$L = 0.410$	$L = 0.062$	$L = 0.291$
	$U = 0.600$	$U = 0.135$	$U = 0.457$

Table 6.26. 95% *intervals for functions of the* μ_j.

Parameter	Lower bound	Upper bound
μ_Y	−40.38	−39.91
μ_1	−40.22	−39.73
μ_2	−40.63	−40.14
μ_3	−40.32	−39.83
$\mu_1 - \mu_2$	0.31	0.52
$\mu_1 - \mu_3$	−0.00015	0.21
$\mu_2 - \mu_3$	−0.42	−0.20

$\overline{y}_{*1*} = -39.972$, $\overline{y}_{*2*} = -40.387$, $\overline{y}_{*3*} = -40.079$, $USL = -33$, $LSL = -41$, and the mean squares in Table 6.23. The value of m used for the intervals on μ_j is 19 (truncated). All the intervals are very similar to those obtained with the interaction model in Section 6.7.

6.10 Summary

The MLS intervals presented in this chapter are simple modifications of the formulas in Chapter 3 (random two-factor model with interaction) and Chapter 5 (random two-factor model with no interaction). In most cases, these modifications provide intervals with confidence coefficients at least as great as the stated level. We have also presented GPQs that generally provide intervals with confidence coefficients close to the stated level.

In the next chapter we consider unbalanced gauge R&R experiments. We consider both random and mixed models and offer recommendations on how to proceed when faced with unequal replicates in the data.

Chapter 7

Unbalanced One- and Two-Factor Models

7.1 Introduction

All the data sets considered to this point have been balanced. A balanced data set has equal replications for each treatment condition. Under the assumptions of normality and independence, all mean squares in a balanced random model are functions of chi-squared random variables. This is an important assumption if the intervals presented in the previous chapters are to maintain the stated confidence levels.

Unfortunately, conditions sometimes arise that result in unbalanced designs. For example, some measurements might be lost due to a gauge malfunction, or an operator might not be able to complete an entire experiment. When a design is unbalanced, some of the distributional properties of the mean squares are lost, and the intervals derived under the balanced model assumptions are no longer valid. In this chapter we describe an approach for constructing confidence intervals with unbalanced data using unweighted sums of squares. We present detailed results for the one-factor and two-factor models discussed previously in this book. References for unbalanced nested designs are provided in Section 8.3.9.

7.2 Unbalanced One-Factor Random Models

We begin by considering the one-factor random model presented in Chapter 2. Table 7.1 reports a modified version of the data in Table 2.1, where only one observation is collected for parts 2, 3, and 7. Recall that these data represent thermal performances (in C°) of a sample of $p = 10$ power modules. The measurements were all taken by a single operator.

The unbalanced one-factor random model is

$$Y_{ij} = \mu_Y + P_i + E_{ij}, \qquad (7.1)$$
$$i = 1, \ldots, p, \quad j = 1, \ldots, r_i,$$

where μ_Y is a constant and P_i and E_{ij} are jointly independent normal random variables with means of zero and variances σ_P^2 and σ_E^2, respectively. The traditional ANOVA for model (7.1) is shown in Table 7.2, where $N = \Sigma_i r_i$, $r_0 = (N - \Sigma_i r_i^2/N)/(p-1)$, and r_i represents the number of replicates on part i. Definitions of mean squares and means are shown in Table 7.3.

Table 7.1. *Example data for unbalanced one-factor random model.*

Part	Measurements
1	37, 38
2	42
3	30
4	42, 43
5	28, 30
6	42, 42
7	25
8	40, 40
9	25, 25
10	35, 34

Table 7.2. *Traditional ANOVA for model* (7.1).

Source of variation	Degrees of freedom	Mean square	Expected mean square
Parts (P)	$p-1$	S_P^2	$\theta_P = \sigma_E^2 + r_0\sigma_P^2$
Replicates (E)	$N-p$	S_E^2	$\theta_E = \sigma_E^2$

Table 7.3. *Mean squares and means used in Table* 7.2.

Statistic	Definition
S_P^2	$\dfrac{\Sigma_i r_i(\overline{Y}_{i*} - \overline{Y}_{**})^2}{p-1}$
S_E^2	$\dfrac{\Sigma_i \Sigma_j (Y_{ij} - \overline{Y}_{i*})^2}{N-p}$
$\overline{Y}_{i*}$	$\dfrac{\Sigma_j Y_{ij}}{r_i}$
$\overline{Y}_{**}$	$\dfrac{\Sigma_i \Sigma_j Y_{ij}}{N}$

As with the balanced model, S_P^2 and S_E^2 are independent, and $(N-p)S_E^2/\theta_E$ has a chi-squared distribution with $N-p$ degrees of freedom. However, unless $\sigma_P^2 = 0$, $(p-1)S_P^2/\theta_P$ no longer has a chi-squared distribution and the confidence interval formulas in Chapter 2 are no longer valid.

One solution to this problem is to construct confidence intervals using an alternative set of statistics. In particular, we use unweighted sums of squares (USS) for this purpose.

Thomas and Hultquist [62] proposed the USS statistic $(p-1)S^2_{P*}/\theta_{P*}$, where

$$S^2_{P*} = \frac{r_H \Sigma_i(\overline{Y}_{i*} - \overline{Y}^*_{**})^2}{p-1},$$

$$\overline{Y}^*_{**} = \Sigma_i \overline{Y}_{i*}/p,$$

$$r_H = \frac{p}{\Sigma_i 1/r_i},$$

and

$$\theta_{P*} = E(S^2_{P*}) = \sigma^2_E + r_H\sigma^2_P. \tag{7.2}$$

Computationally, $\Sigma_i(\overline{Y}_{i*} - \overline{Y}^*_{**})^2 = \Sigma_i\overline{Y}^2_{i*} - (\Sigma_i\overline{Y}_{i*})^2/p$. The statistic $(p-1)S^2_{P*}$ is the USS of the treatment means and r_H is the harmonic mean of the r_i values.

Thomas and Hultquist [62] showed that under the assumptions of model (7.1), the moment generating function of $(p-1)S^2_{P*}/\theta_{P*}$ approaches that of a chi-squared random variable with $p-1$ degrees of freedom as all r_i approach a constant, or if $\rho = \sigma^2_P/(\sigma^2_P + \sigma^2_E)$ approaches one, or if all r_i approach infinity. Furthermore, they provide simulations that show this approximation works well unless $\rho < 0.20$ and the design is extremely unbalanced.

The condition that $\rho < 0.20$ is not typical in a gauge R&R study since this implies SNR < 0.707. A measurement system with an SNR this low is clearly of no value. Additionally, it would be unusual to have an extremely unbalanced data set in a gauge R&R study. An experiment that would cause the Thomas–Hultquist approximation to perform poorly would include a few parts with only one replication and the remaining parts with a very large number of replications (say, 100 or more). This is unlikely in a gauge R&R study since replicates typically range between one and four. Thus, the Thomas–Hultquist approximation should perform well under the unbalanced conditions typically encountered in a gauge R&R study.

Very simply then, the MLS formulas in Chapter 2 can be modified to account for unbalancedness by replacing S^2_P, r, and $\overline{Y}_{**}$ with S^2_{P*}, r_H, and $\overline{Y}^*_{**}$, respectively. We refer to this approach as the USS modification. Table 7.4 gives the gauge R&R parameters for model (7.1) and the point estimators based on the USS modification. MLS intervals based

Table 7.4. *Gauge R&R parameters and USS point estimators for model* (7.1).

Gauge R&R notation	Model (7.1) representation	Point estimator
μ_Y	μ_Y	$\widehat{\mu}_Y = \overline{Y}^*_{**}$
γ_P	σ^2_P	$\widehat{\gamma}_P = \dfrac{S^2_{P*} - S^2_E}{r_H}$
γ_M	σ^2_E	$\widehat{\gamma}_M = S^2_E$
γ_R	$\dfrac{\sigma^2_P}{\sigma^2_E}$	$\widehat{\gamma}_R = \dfrac{S^2_{P*}/S^2_E - 1}{r_H}$

Table 7.5. *Constants used in confidence intervals for model* (7.1). *Values are for* $\alpha = 0.05$, $p = 10$, *and* $N = 17$.

Constant	Definition	Value
G_1	$1 - F_{\alpha/2:\infty,p-1}$	0.5269
G_2	$1 - F_{\alpha/2:\infty,N-p}$	0.5628
H_1	$F_{1-\alpha/2:\infty,p-1} - 1$	2.333
H_2	$F_{1-\alpha/2:\infty,N-p} - 1$	3.142
F_1	$F_{1-\alpha/2:p-1,N-p}$	4.823
F_2	$F_{\alpha/2:p-1,N-p}$	0.2383
G_{12}	$\dfrac{(F_1-1)^2 - G_1^2F_1^2 - H_2^2}{F_1}$	−0.3556
H_{12}	$\dfrac{(1-F_2)^2 - H_1^2F_2^2 - G_2^2}{F_2}$	−0.1910

on the USS modification follow in this section. Constants used in the formulas are given in Table 7.5.

7.2.1 Interval for μ_Y

The confidence interval for μ_Y is obtained by substituting S^2_{P*} for S^2_P, r_H for r, and $\overline{Y}^*_{**}$ for $\overline{Y}_{**}$ in Equation (2.2). This provides the approximate $100(1-\alpha)\%$ confidence interval for μ_Y,

$$L = \overline{Y}^*_{**} - \sqrt{\frac{S^2_{P*}F_{1-\alpha:1,p-1}}{pr_H}}$$

and

$$U = \overline{Y}^*_{**} + \sqrt{\frac{S^2_{P*}F_{1-\alpha:1,p-1}}{pr_H}}. \tag{7.3}$$

El-Bassiouni and Abdelhafez [21] compared Equation (7.3) to nine other intervals for μ_Y. (The form of Equation (7.3) considered by El-Bassiouni and Abdelhafez used an estimated degrees of freedom based on the Satterthwaite approximation. However, they found this was adequately approximated by $p-1$.) Equation (7.3) maintained the stated confidence level and provided shorter intervals than other methods considered in the study. Additionally, as ρ approaches one, the confidence coefficient of Equation (7.3) seems to converge to the stated level. The only situation in which Equation (7.3) provided wider intervals than some of the alternatives was when ρ was small. As noted earlier, such a condition is not typical in a gauge R&R study.

7.2.2 Interval for γ_P

An approximate $100(1-\alpha)\%$ confidence interval for $\gamma_P = \sigma_P^2$ based on the USS modification of Equation (2.3) is

$$L = \widehat{\gamma}_P - \frac{\sqrt{V_{LP}}}{r_H}$$

and

$$U = \widehat{\gamma}_P + \frac{\sqrt{V_{UP}}}{r_H}, \quad (7.4)$$

where

$$V_{LP} = G_1^2 S_{P*}^4 + H_2^2 S_E^4 + G_{12} S_{P*}^2 S_E^2,$$
$$V_{UP} = H_1^2 S_{P*}^4 + G_2^2 S_E^4 + H_{12} S_{P*}^2 S_E^2,$$

G_1, G_2, H_1, H_2, G_{12}, and H_{12} are defined in Table 7.5 and $\widehat{\gamma}_P$ is defined in Table 7.4.

A similar modification of Equation (2.4) provides the approximate $100(1-\alpha)\%$ confidence interval for γ_P,

$$L = \frac{(S_{P*}^2 - S_E^2 F_1)(1 - G_1)}{r_H}$$

and

$$U = \frac{(S_{P*}^2 - S_E^2 F_2)(1 + H_1)}{r_H}, \quad (7.5)$$

where G_1, H_1, F_1, and F_2 are defined in Table 7.5.

Lee and Khuri [40] used simulation to compare confidence coefficients and average intervals lengths of Equation (7.5) with three other intervals. (They did not include Equation (7.4) in their study.) They concluded that Equation (7.5) can be recommended unless the design is very unbalanced and ρ is very small. Our own investigations have shown that Equations (7.4) and (7.5) provide comparable intervals.

If an investigator encounters a situation in which the data are very unbalanced and ρ is very small, El-Bassiouni [20] recommends a computationally intensive method based on the exact interval for ρ proposed by Wald [68, 69]. Alternatively, Burdick and Eickman [9] proposed a closed-form interval that is guaranteed to provide a confidence coefficient at least as great as the stated level. This approximate $100(1-\alpha)\%$ interval for γ_P is

$$L = \frac{S_{P*}^2 L_m (1 - G_1)}{1 + r_H L_m}$$

and

$$U = \frac{S_{P*}^2 U_M (1 + H_1)}{1 + r_H U_M}, \quad (7.6)$$

where

$$L_m = \frac{S_{P*}^2}{r_H S_E^2 F_1} - \frac{1}{m},$$

$$U_M = \frac{S_{P*}^2}{r_H S_E^2 F_2} - \frac{1}{M},$$

$$m = \text{Min}(r_1, r_2, \ldots, r_P),$$

$$M = \text{Max}(r_1, r_2, \ldots, r_P), \tag{7.7}$$

and G_1 and H_1 are defined in Table 7.5. Note that Equations (7.5) and (7.6) both simplify to Equation (2.4) when all r_i are equal.

To summarize, we recommend either Equation (7.4) or (7.5) for a typical gauge R&R application. Both intervals will maintain the stated confidence level and provide shorter intervals than Equation (7.6).

7.2.3 Interval for γ_M

The exact $100(1-\alpha)\%$ confidence interval for $\gamma_M = \sigma_E^2$ is

$$L = (1 - G_2)S_E^2$$

and

$$U = (1 + H_2)S_E^2, \tag{7.8}$$

where G_2 and H_2 are defined in Table 7.5. This is the same interval presented in Equation (2.5) since $(N - p)S_E^2/\theta_E$ has a chi-squared distribution in both the balanced and the unbalanced model.

7.2.4 Interval for γ_R

The approximate $100(1-\alpha)\%$ confidence interval for γ_R formed by making the USS modification to Equation (2.6) is

$$L = \frac{S_{P*}^2}{r_H S_E^2 F_1} - \frac{1}{r_H}$$

and

$$U = \frac{S_{P*}^2}{r_H S_E^2 F_2} - \frac{1}{r_H}. \tag{7.9}$$

Equation (7.9) provides an interval that generally meets the stated confidence level unless ρ is small. Donner and Wells [19] reported simulation results for this interval with $p = 25$ and $p = 50$. The simulated confidence coefficient for a 95% two-sided interval was below the stated level in both designs for $\rho = 0$ and 0.1 and also in the design with $p = 25$ for $\rho = 0.2$.

In an application where ρ is small, one can compute an exact interval for γ_R using an exact confidence interval for ρ proposed by Wald [68, 69]. (An interval for ρ can be transformed to an interval for γ_R using the relation in Equation (1.4).) Computation of this interval requires solution of two nonlinear equations. Harville and Fenech [28] provided an iterative solution technique for these equations that is consistent with an approach proposed by Seely and El-Bassiouni [59]. Lee and Seely [41] provided an alternative solution technique that employs Newton's method, and Burdick, Maqsood, and Graybill [12] used the bisection method. SAS code for the latter approach is provided in Appendix B of [10].

Burdick, Maqsood, and Graybill [12] proposed an MLS interval for γ_R that provides lower and upper confidence bounds with confidence of at least $100(1-\alpha)\%$. This interval is

$$L = L_m$$

and

$$U = U_M, \tag{7.10}$$

where L_m and U_M are defined in Equation (7.7). Equation (7.10) can be very conservative if the design is very unbalanced or ρ is very small. In these conditions, exact solution of the Wald interval is recommended. Alternatives to Wald's interval have been investigated by Lin and Harville [43], LaMotte, McWhorter, and Prasad [38], and Burch and Iyer [5].

In summary, we recommend Equation (7.9) for constructing a confidence interval on γ_R in a typical gauge R&R application. This interval will generally maintain the stated confidence level and is easier to compute than the exact Wald interval. In applications where ρ is small, Equation (7.10) provides confidence intervals with confidence coefficients at least as great as the stated level. Computation of the exact Wald interval is also recommended in this case, particularly if the design is extremely unbalanced.

7.2.5 GCIs

The USS modification can be used to construct a GCI for γ_P. In particular, one can use the GPQ shown in Equation (2.7) after replacing s_P^2 and r with s_{P*}^2 and r_H, respectively. The degrees of freedom for W_2 become $N-p$ instead of $p(r-1)$. The computed intervals will be very similar to the MLS intervals.

Park and Burdick [56] proposed an approach for constructing a generalized interval for γ_P based on the results of Olsen, Seely, and Birkes [54]. The method employs exact chi-squared random variables and requires solution of a nonlinear equation. Details are provided in Section 7.6.

7.2.6 Intervals for Misclassification Rates

We again use generalized inference to compute confidence intervals for the misclassification rates. Since ρ is expected to be large in most gauge R&R studies, we recommend using the algorithm described in Section 2.4 after applying the USS modification to the GPQs in Table 2.7.

Table 7.6. *Example data for unbalanced one-factor random model.*

Part	Measurements	Part average ($\overline{Y}_{i*}$)
1	37, 38	37.5
2	42	42
3	30	30
4	42, 43	42.5
5	28, 30	29
6	42, 42	42
7	25	25
8	40, 40	40
9	25, 25	25
10	35, 34	34.5

7.2.7 Numerical Example

We now show how to compute the confidence intervals using the data in Table 7.1. For convenience these data are repeated in Table 7.6. The first step is to compute the USS statistics S^2_{P*}, r_H, and $\overline{Y}^*_{**}$ defined in Equation (7.2).

To compute these values, perform the following steps:

1. Compute the sample average for each part. The computed values for our example are shown in the last column of Table 7.6.

2. Compute $\Sigma_i(\overline{Y}_{i*} - \overline{Y}^*_{**})^2$ using the averages computed in step 1. For our example,

$$\begin{aligned}\Sigma_i(\overline{Y}_{i*} - \overline{Y}^*_{**})^2 &= \Sigma_i\overline{Y}^2_{i*} - \frac{(\Sigma_i\overline{Y}_{i*})^2}{p}\\ &= (37.5^2 + \cdots + 34.5^2) - \frac{(37.5 + \cdots + 34.5)^2}{10}\\ &= (12,521.75) - \frac{(347.5)^2}{10}\\ &= 446.125.\end{aligned}$$

3. Compute $\overline{Y}^*_{**}$ using the averages from step 1. Here,

$$\begin{aligned}\overline{Y}^*_{**} &= \frac{\Sigma_i\overline{Y}_{i*}}{p}\\ &= \frac{37.5 + 42 + \cdots + 34.5}{10}\\ &= 34.75.\end{aligned}$$

4. Compute r_H. For our example,

$$r_H = \frac{p}{\Sigma_i 1/r_i} = \frac{10}{1/2 + 1/1 + \cdots + 1/2} = \frac{10}{6.5} = 1.53846.$$

5. Compute S^2_{P*} using results from steps 2 and 4. For our example,

$$S^2_{P*} = \frac{r_H \Sigma_i (\overline{Y}_{i*} - \overline{Y}^*_{**})^2}{p-1} = \frac{1.53846(446.125)}{10-1} = 76.26.$$

6. Compute S^2_E using the individual measurements and the averages from step 1. For our example,

$$S^2_E = \frac{\Sigma_i \Sigma_j (Y_{ij} - \overline{Y}_{i*})^2}{N-p} = \frac{(37-37.5)^2 + (38-37.5)^2 + \cdots + (34-34.5)^2}{17-10} = 0.5.$$

We now compute confidence intervals for gauge R&R parameters and the misclassification rates. The constants used in the computations are shown in Table 7.5.

The 95% confidence interval for μ_Y using Equation (7.3) is

$$L = \overline{Y}^*_{**} - \sqrt{\frac{S^2_{P*} F_{1-\alpha:1,p-1}}{pr_H}} = 34.75 - \sqrt{\frac{76.26(5.1174)}{10(1.53846)}} = 29.7$$

and

$$U = \overline{Y}^*_{**} + \sqrt{\frac{S^2_{P*} F_{1-\alpha:1,p-1}}{pr_H}} = 34.75 + \sqrt{\frac{76.26(5.1174)}{10(1.53846)}} = 39.8.$$

An approximate 95% confidence interval for γ_P using Equation (7.4) requires

$$\begin{aligned} V_{LP} &= G_1^2 S_{P*}^4 + H_2^2 S_E^4 + G_{12} S_{P*}^2 S_E^2 \\ &= (0.5269)^2(76.26)^2 + (3.142)^2(0.5)^2 + (-0.3556)(76.26)(0.5) \\ &= 1.60 \times 10^3 \end{aligned}$$

and

$$\begin{aligned} V_{UP} &= H_1^2 S_{P*}^4 + G_2^2 S_E^4 + H_{12} S_{P*}^2 S_E^2 \\ &= (2.333)^2(76.26)^2 + (0.5628)^2(0.5)^2 + (-0.1910)(76.26)(0.5) \\ &= 3.16 \times 10^4. \end{aligned}$$

The computed 95% confidence interval for γ_P is then

$$\begin{aligned} L &= \widehat{\gamma}_P - \frac{\sqrt{V_{LP}}}{r_H} \\ &= \frac{76.26 - 0.5}{1.53846} - \frac{\sqrt{1.60 \times 10^3}}{1.53846} \\ &= 23.2 \end{aligned}$$

and

$$\begin{aligned} U &= \widehat{\gamma}_P + \frac{\sqrt{V_{UP}}}{r_H} \\ &= \frac{76.26 - 0.5}{1.53846} + \frac{\sqrt{3.16 \times 10^4}}{1.53846} \\ &= 165. \end{aligned} \tag{7.11}$$

The 95% confidence interval for γ_P using Equation (7.5) is

$$\begin{aligned} L &= \frac{(S_{P*}^2 - S_E^2 F_1)(1 - G_1)}{r_H} \\ &= \frac{[76.26 - 0.5(4.823)](1 - 0.5269)}{1.53846} \\ &= 22.7 \end{aligned}$$

and

$$\begin{aligned} U &= \frac{(S_{P*}^2 - S_E^2 F_2)(1 + H_1)}{r_H} \\ &= \frac{[76.26 - 0.5(0.2383)](1 + 2.333)}{1.53846} \\ &= 165. \end{aligned}$$

Using Equations (7.6) and (7.7) to compute a 95% confidence interval for γ_P, we first obtain $m = 1$, $M = 2$,

$$\begin{aligned} L_m &= \frac{S^2_{P*}}{r_H S^2_E F_1} - \frac{1}{m} \\ &= \frac{76.26}{1.53846(0.5)(4.823)} - \frac{1}{1} \\ &= 19.55 \end{aligned}$$

and

$$\begin{aligned} U_M &= \frac{S^2_{P*}}{r_H S^2_E F_2} - \frac{1}{M} \\ &= \frac{76.26}{1.53846(0.5)(0.2383)} - \frac{1}{2} \\ &= 415.5. \end{aligned} \tag{7.12}$$

This yields the 95% interval for γ_P,

$$\begin{aligned} L &= \frac{S^2_{P*} L_m (1 - G_1)}{1 + r_H L_m} \\ &= \frac{76.26(19.55)(1 - 0.5269)}{1 + 1.53846(19.55)} \\ &= 22.7 \end{aligned}$$

and

$$\begin{aligned} U &= \frac{S^2_{P*} U_M (1 + H_1)}{1 + r_H U_M} \\ &= \frac{76.26(415.5)(1 + 2.333)}{1 + 1.53846(415.5)} \\ &= 165. \end{aligned}$$

The exact 95% confidence interval for γ_M computed from Equation (7.8) is

$$\begin{aligned} L &= (1 - G_2) S^2_E \\ &= (1 - 0.5628)(0.5) \\ &= 0.219 \end{aligned}$$

and

$$\begin{aligned} U &= (1 + H_2) S^2_E \\ &= (1 + 3.142)(0.5) \\ &= 2.07. \end{aligned} \tag{7.13}$$

The 95% confidence interval for γ_R using Equation (7.9) is

$$\begin{aligned} L &= \frac{S_{P*}^2}{r_H S_E^2 F_1} - \frac{1}{r_H} \\ &= \frac{76.26}{1.53846(0.5)(4.823)} - \frac{1}{1.53846} \\ &= 19.9 \end{aligned}$$

and

$$\begin{aligned} U &= \frac{S_{P*}^2}{r_H S_E^2 F_2} - \frac{1}{r_H} \\ &= \frac{76.26}{1.53846(0.5)(0.2383)} - \frac{1}{1.53846} \\ &= 415. \end{aligned} \tag{7.14}$$

Computation of an interval for γ_R using Equation (7.10) requires L_m and U_M previously computed in Equation (7.12). This provides the 95% interval $L = 19.6$ and $U = 416$. For comparison, the exact 95% Wald interval computed as described in Appendix B of [10] is $L = 19.9$ and $U = 416$. The intervals in this example are so similar because ρ appears to be very close to one.

Using Equation (1.2) with $k = 6$, $LSL = 18$, and $USL = 58$, a 95% confidence interval for PTR based on the interval calculated for γ_M in Equation (7.13) is

$$L = \frac{6\sqrt{0.219}}{58 - 18} = 0.070$$

and

$$U = \frac{6\sqrt{2.07}}{58 - 18} = 0.216.$$

A 95% confidence interval for SNR based on the computed confidence interval for γ_R in Equation (7.14) is

$$L = \sqrt{2(19.9)} = 6.31$$

and

$$U = \sqrt{2(415)} = 28.8.$$

The 95% confidence interval for C_p based on the computed interval for γ_P in Equation (7.11) is

$$L = \frac{58 - 18}{6\sqrt{165}} = 0.519$$

and

$$U = \frac{58 - 18}{6\sqrt{23.2}} = 1.38.$$

Table 7.7 gives confidence intervals for the misclassification rates. These rates are based on a simulation of 100,000 GPQ values with $\overline{y}_{**}^* = 34.75$ and $\epsilon = 0.001$. Both δ_{index} and β_{index} have upper bounds less than one.

Table 7.7. *95% confidence intervals for misclassification rates. (Bounds for δ, δ_c, β, and β_c have been multiplied by 10^6.)*

Parameter	Lower bound	Upper bound
δ	57	8,351
δ_c	57	9,733
δ_{index}	0.044	0.499
β	28	6,993
β_c	34,498	205,613
β_{index}	0.040	0.206

Table 7.8. *Summary of USS 95% intervals for one-factor example. (See Section 1.8 for a description of computer programs to perform these computations.)*

Parameter	Lower bound	Upper bound
μ_Y	29.7	39.8
γ_P	23.2	165
γ_M	0.219	2.07
γ_R	19.9	415
PTR	0.070	0.216
SNR	6.31	28.8
C_p	0.519	1.38

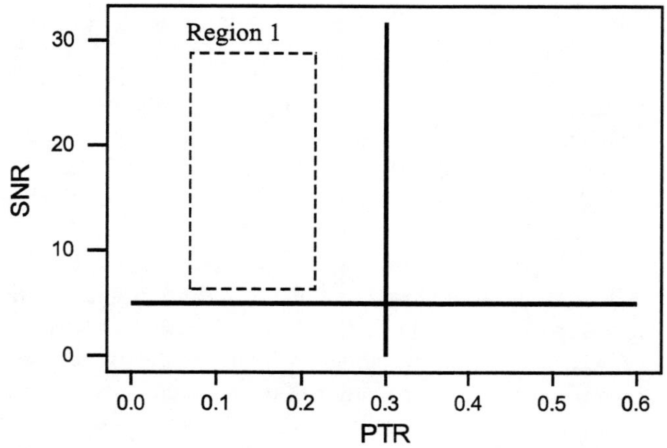

Figure 7.1. *R&R graph for unbalanced one-factor example.*

Table 7.8 gives the computed intervals for this example. The R&R graph in Figure 7.1 suggests the measurement system satisfies both the PTR and SNR criteria. The residual plots in Figure 7.2 do not suggest any assumption violations.

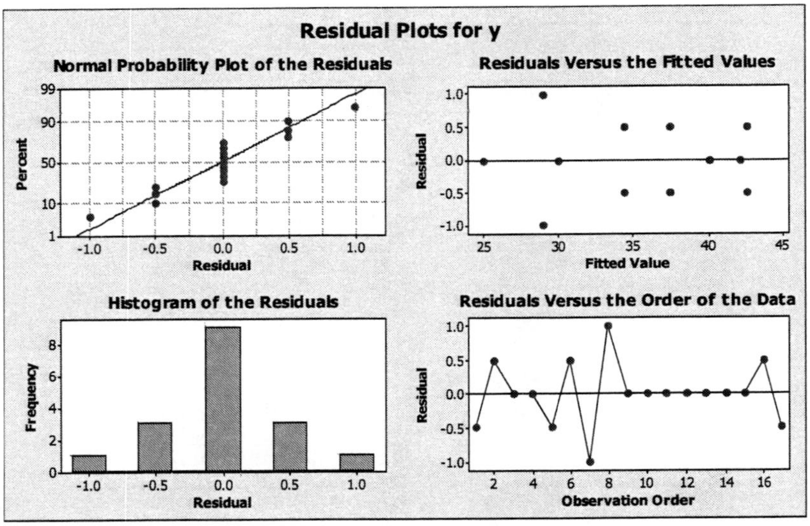

Figure 7.2. *Residual plots for the one-factor unbalanced example.*

7.3 Unbalanced Two-Factor Random Model with Interaction

The unbalanced two-factor random model with interaction is

$$Y_{ijk} = \mu_Y + P_i + O_j + (PO)_{ij} + E_{ijk}, \qquad (7.15)$$
$$i = 1, \ldots, p, \quad j = 1, \ldots, o, \quad k = 1, \ldots, r_{ij} > 0,$$

where μ_Y is a constant and P_i, O_j, $(PO)_{ij}$, and E_{ijk} are jointly independent normal random variables with means of zero and variances $\sigma_P^2, \sigma_O^2, \sigma_{PO}^2$, and σ_E^2, respectively. The term r_{ij} represents the number of replicates in cell (i, j). In the context of a gauge R&R study, r_{ij} is the number of times part i is measured by operator j. The total number of observations is $N = \Sigma_i \Sigma_j r_{ij}$.

The confidence interval formulas for the balanced two-factor model in Chapter 3 are based on the assumption that the mean squares are independent and have chi-squared distributions. However, for the unbalanced model (7.15), there is no set of independent mean squares that have exact chi-squared distributions. Thus, we again use a set of USS estimators that approximate the distributional properties of the balanced model mean squares.

Hernandez and Burdick [29] used the USS modification to construct good approximate confidence intervals for σ_P^2 and σ_O^2 in model (7.15). Gong, Burdick, and Quiroz [22] extended these results to construct confidence intervals for the gauge R&R parameters shown in Table 7.9.

The USS ANOVA for model (7.15) is shown in Table 7.10. Definitions for means and mean squares are shown in Table 7.11. Comparison of the expected mean squares with those in Table 3.2 shows the only difference is that r is replaced with r_H. Under the assumptions of model (7.15), $(p-1)S_{P*}^2/\theta_{P*}$, $(o-1)S_{O*}^2/\theta_{O*}$, and $(p-1)(o-1)S_{PO*}^2/\theta_{PO*}$ have

Table 7.9. *Gauge R&R parameters and point estimators for model* (7.15).

Gauge R&R notation	Model (7.15) representation	USS point estimator
μ_Y	μ_Y	$\widehat{\mu}_Y = \overline{Y}^*_{***}$
γ_P	σ^2_P	$\widehat{\gamma}_P = \dfrac{S^2_{P*} - S^2_{PO*}}{or_H}$
γ_M	$\sigma^2_O + \sigma^2_{PO} + \sigma^2_E$	$\widehat{\gamma}_M = \dfrac{S^2_{O*} + (p-1)S^2_{PO*} + p(r_H - 1)S^2_E}{pr_H}$
γ_R	$\dfrac{\sigma^2_P}{\sigma^2_O + \sigma^2_{PO} + \sigma^2_E}$	$\widehat{\gamma}_R = \dfrac{\widehat{\gamma}_P}{\widehat{\gamma}_M}$

Table 7.10. *USS ANOVA for model* (7.15).

Source of variation	Degrees of freedom	Mean square	Expected mean square
Parts (P)	$p-1$	S^2_{P*}	$\theta_{P*} = \sigma^2_E + r_H\sigma^2_{PO} + or_H\sigma^2_P$
Operators (O)	$o-1$	S^2_{O*}	$\theta_{O*} = \sigma^2_E + r_H\sigma^2_{PO} + pr_H\sigma^2_O$
P×O	$(p-1)(o-1)$	S^2_{PO*}	$\theta_{PO*} = \sigma^2_E + r_H\sigma^2_{PO}$
Replicates	$N - po$	S^2_E	$\theta_E = \sigma^2_E$

approximate chi-squared distributions with $(p-1)$, $(o-1)$, and $(p-1)(o-1)$ degrees of freedom, respectively. The random variable $(N - po)S^2_E/\theta_E$ has an exact chi-squared distribution with $(N - po)$ degrees of freedom.

7.3.1 Confidence Intervals for Gauge R&R Parameters

Confidence intervals in the unbalanced two-factor model can be computed using the balanced model formulas with the USS modification. This means replacing r with r_H, S^2_P with S^2_{P*}, S^2_O with S^2_{O*}, and S^2_{PO} with S^2_{PO*} in the formulas from Chapter 3. Table 7.12 presents a summary of the resulting gauge R&R intervals. Terms used in Table 7.12 are defined in Table 7.13, and the constants are defined in Table 7.14.

It must be noted that the USS modification can be used only if there are no missing cells (that is, all $r_{ij} > 0$). Although some authors have suggested alternative sums of squares for designs with missing cells (see, e.g., Kazempour and Graybill [36]), it might be more practical to simply discard all measurements for any part that is not measured by all operators. Given that most gauge R&R studies have a large number of parts, this should not have a major impact on the analysis. GCIs based on the USS modification of the GPQs in Table 3.8 will provide results similar to the intervals in Table 7.12.

Table 7.11. *USS mean squares and means for model* (7.15).

Statistic	Definition
S^2_{P*}	$\dfrac{or_H \Sigma_i (\overline{Y}^*_{i**} - \overline{Y}^*_{***})^2}{p-1}$
S^2_{O*}	$\dfrac{pr_H \Sigma_j (\overline{Y}^*_{*j*} - \overline{Y}^*_{***})^2}{o-1}$
S^2_{PO*}	$\dfrac{r_H \Sigma_i \Sigma_j (\overline{Y}_{ij*} - \overline{Y}^*_{i**} - \overline{Y}^*_{*j*} + \overline{Y}^*_{***})^2}{(p-1)(o-1)}$
S^2_E	$\dfrac{\Sigma_i \Sigma_j \Sigma_k (Y_{ijk} - \overline{Y}_{ij*})^2}{N - po}$
$\overline{Y}_{ij*}$	$\dfrac{\Sigma_k Y_{ijk}}{r_{ij}}$
$\overline{Y}^*_{i**}$	$\dfrac{\Sigma_j \overline{Y}_{ij*}}{o}$
$\overline{Y}^*_{*j*}$	$\dfrac{\Sigma_i \overline{Y}_{ij*}}{p}$
$\overline{Y}^*_{***}$	$\dfrac{\Sigma_i \Sigma_j \overline{Y}_{ij*}}{po}$
r_H	$\dfrac{po}{\Sigma_i \Sigma_j 1/r_{ij}}$

Table 7.12. *Confidence intervals for gauge R&R parameters in model* (7.15).

Parameter	Balanced equation	USS modification Lower bound	USS modification Upper bound
μ_Y	(3.2)	$\overline{Y}^*_{***} - C\sqrt{\frac{K}{por_H}}$	$\overline{Y}^*_{***} + C\sqrt{\frac{K}{por_H}}$
γ_P	(3.3)	$\widehat{\gamma}_P - \dfrac{\sqrt{V_{LP}}}{or_H}$	$\widehat{\gamma}_P + \dfrac{\sqrt{V_{UP}}}{or_H}$
γ_M	(3.4)	$\widehat{\gamma}_M - \dfrac{\sqrt{V_{LM}}}{pr_H}$	$\widehat{\gamma}_M + \dfrac{\sqrt{V_{UM}}}{pr_H}$
γ_R	(3.5)	$\dfrac{p(1-G_1)(S^2_{P*} - F_1 S^2_{PO*})}{V_{LR}}$	$\dfrac{p(1+H_1)(S^2_{P*} - F_2 S^2_{PO*})}{V_{UR}}$

Table 7.13. *Terms used in Table* 7.12.

Term	Definition
K	$S^2_{P*} + S^2_{O*} - S^2_{PO*}$
C	$\dfrac{S^2_{P*}\sqrt{F_{1-\alpha:1,p-1}} + S^2_{O*}\sqrt{F_{1-\alpha:1,o-1}} - S^2_{PO*}\sqrt{F_{1-\alpha:1,(p-1)(o-1)}}}{K}$
$\widehat{\gamma}_P$	$\dfrac{S^2_{P*} - S^2_{PO*}}{or_H}$
V_{LP}	$G_1^2 S^4_{P*} + H_3^2 S^4_{PO*} + G_{13} S^2_{P*} S^2_{PO*}$
V_{UP}	$H_1^2 S^4_{P*} + G_3^2 S^4_{PO*} + H_{13} S^2_{P*} S^2_{PO*}$
$\widehat{\gamma}_M$	$\dfrac{S^2_{O*} + (p-1)S^2_{PO*} + p(r_H - 1)S^2_E}{pr_H}$
V_{LM}	$G_2^2 S^4_{O*} + G_3^2(p-1)^2 S^4_{PO*} + G_4^2 p^2 (r_H - 1)^2 S^4_E$
V_{UM}	$H_2^2 S^4_{O*} + H_3^2(p-1)^2 S^4_{PO*} + H_4^2 p^2 (r_H - 1)^2 S^4_E$
V_{LR}	$po(r_H - 1)S^2_E + o(1 - G_1)F_3 S^2_{O*} + o(p-1)S^2_{PO*}$
V_{UR}	$po(r_H - 1)S^2_E + o(1 + H_1)F_4 S^2_{O*} + o(p-1)S^2_{PO*}$

Table 7.14. *Constants used in Table* 7.12. *Values are for* $\alpha = 0.05$, $p = 10$, $o = 3$, *and* $N = 83$.

Constant	Definition	Value
G_1	$1 - F_{\alpha/2:\infty,p-1}$	0.5269
G_2	$1 - F_{\alpha/2:\infty,o-1}$	0.7289
G_3	$1 - F_{\alpha/2:\infty,(p-1)(o-1)}$	0.4290
G_4	$1 - F_{\alpha/2:\infty,N-po}$	0.2934
H_1	$F_{1-\alpha/2:\infty,p-1} - 1$	2.333
H_2	$F_{1-\alpha/2:\infty,o-1} - 1$	38.50
H_3	$F_{1-\alpha/2:\infty,(p-1)(o-1)} - 1$	1.187
H_4	$F_{1-\alpha/2:\infty,N-po} - 1$	0.5240
F_1	$F_{1-\alpha/2:p-1,(p-1)(o-1)}$	2.929
F_2	$F_{\alpha/2:p-1,(p-1)(o-1)}$	0.2702
F_3	$F_{1-\alpha/2:p-1,o-1}$	39.39
F_4	$F_{\alpha/2:p-1,o-1}$	0.1750
G_{13}	$\dfrac{(F_1 - 1)^2 - G_1^2 F_1^2 - H_3^2}{F_1}$	−0.02358
H_{13}	$\dfrac{(1 - F_2)^2 - H_1^2 F_2^2 - G_3^2}{F_2}$	−0.1800

Table 7.15. *Data for unbalanced two-factor example.*

Part	Operator 1	Operator 2	Operator 3
1	37, 38	41, 41, 40	41, 42, 41
2	42, 41, 43	42, 42, 42	43
3	30, 31, 31	31, 31, 31	29, 30, 28
4	42, 43, 42	43, 43, 43	42, 42, 42
5	28, 30, 29	29, 30, 29	31, 29, 29
6	42, 42, 43	45, 45, 45	44, 46, 45
7	25, 26, 27	28, 30	29, 27, 27
8	40, 40, 40	43, 42, 42	43, 43
9	25, 25, 25	27, 29, 28	26, 26, 26
10	35, 34, 34	35, 35, 34	35

7.3.2 Intervals for Misclassification Rates

GCIs for the misclassification rates are computed using the algorithm described in Section 3.4 after applying the USS modification to the GPQs shown in Table 3.9.

7.3.3 Numerical Example

We now compute intervals for the data shown in Table 7.15 using the formulas defined in Table 7.12. This is an artificial data set generated from Table 3.1 with unequal replications in some of the cells. We selected only 10 parts and 3 operators to more easily demonstrate the hand calculations. As noted in Section 4.3, we do not recommend using only 10 parts and 3 operators in a random two-factor design. There are $N = 83$ observations in the data set.

The first step is to compute the USS statistics defined in Table 7.11. The cell means $(\overline{Y}_{ij*})$, the unweighted means for parts $(\overline{Y}^*_{i**})$, the unweighted means for operators $(\overline{Y}^*_{*j*})$, and the unweighted overall mean $(\overline{Y}^*_{***})$ are shown in Table 7.16. To compute the USS mean squares, perform the following steps:

1. Compute r_H. For our example,

$$
\begin{aligned}
r_H &= \frac{po}{\Sigma_i \Sigma_j 1/r_{ij}} \\
&= \frac{10(3)}{1/2 + 1/3 + \cdots + 1/1} \\
&= \frac{30}{11.8333} \\
&= 2.5352.
\end{aligned}
$$

Table 7.16. *Cell means and unweighted means for example.*

Part	Operator 1	Operator 2	Operator 3	$\overline{Y}^*_{i**}$
1	37.5	40.6667	41.3333	39.8333
2	42	42	43	42.3333
3	30.6667	31	29	30.2222
4	42.3333	43	42	42.4444
5	29	29.3333	29.6667	29.3333
6	42.3333	45	45	44.1111
7	26	29	27.6667	27.5556
8	40	42.3333	43	41.7778
9	25	28	26	26.3333
10	34.3333	34.6667	35	34.6667
$\overline{Y}^*_{*j*}$	34.9167	36.5	36.1667	$\overline{Y}^*_{***} = 35.8611$

2. Compute $\Sigma_i(\overline{Y}^*_{i**} - \overline{Y}^*_{***})^2$ using the means in Table 7.16. For our example,

$$\Sigma_i(\overline{Y}^*_{i**} - \overline{Y}^*_{***})^2 = \Sigma_i \overline{Y}^{*2}_{i**} - \frac{(\Sigma_i \overline{Y}^*_{i**})^2}{p}$$

$$= 13,299.867 - \frac{(358.6111)^2}{10}$$

$$= 439.67.$$

3. Compute S^2_{P*} using results from steps 1 and 2. For our example,

$$S^2_{P*} = \frac{or_H \Sigma_i(\overline{Y}^*_{i**} - \overline{Y}^*_{***})^2}{p-1}$$

$$= \frac{3(2.5352)(439.67)}{9}$$

$$= 371.6.$$

4. Compute $\Sigma_j(\overline{Y}^*_{*j*} - \overline{Y}^*_{***})^2$ using the means in Table 7.16. For our example,

$$\Sigma_j(\overline{Y}^*_{*j*} - \overline{Y}^*_{***})^2 = \Sigma_j \overline{Y}^{*2}_{*j*} - \frac{(\Sigma_j \overline{Y}^*_{*j*})^2}{o}$$

$$= (3,859.4514) - \frac{(107.5833)^2}{3}$$

$$= 1.3935.$$

5. Compute S^2_{O*} using results from steps 1 and 4. For our example,

$$\begin{aligned} S^2_{O*} &= \frac{pr_H \Sigma_j (\overline{Y}^*_{*j*} - \overline{Y}^*_{***})^2}{o-1} \\ &= \frac{10(2.5352)(1.3935)}{2} \\ &= 17.66. \end{aligned}$$

6. Compute $\Sigma_i \Sigma_j (\overline{Y}_{ij*} - \overline{Y}^*_{i**} - \overline{Y}^*_{*j*} + \overline{Y}^*_{***})^2$ using the means in Table 7.16. For our example,

$$\begin{aligned} \Sigma_i \Sigma_j (\overline{Y}_{ij*} - \overline{Y}^*_{i**} - \overline{Y}^*_{*j*} + \overline{Y}^*_{***})^2 &= \Sigma_i \Sigma_j \overline{Y}^2_{ij*} - o\Sigma_i \overline{Y}^{*2}_{i**} - p\Sigma_j \overline{Y}^{*2}_{*j*} \\ &\quad + \frac{(\Sigma_i \Sigma_j \overline{Y}_{ij*})^2}{po} \\ &= 39{,}930.806 - 39{,}899.602 \\ &\quad - 38{,}594.514 + 38{,}580.579 \\ &= 17.269. \end{aligned}$$

7. Compute S^2_{PO*} using the results in steps 1 and 6. For our example,

$$\begin{aligned} S^2_{PO*} &= \frac{r_H \Sigma_i \Sigma_j (\overline{Y}_{ij*} - \overline{Y}^*_{i**} - \overline{Y}^*_{*j*} + \overline{Y}^*_{***})^2}{(p-1)(o-1)} \\ &= \frac{2.5352(17.269)}{18} \\ &= 2.432. \end{aligned}$$

8. Finally, compute S^2_E using the individual measurements and the cell means in Table 7.16. For our example,

$$\begin{aligned} S^2_E &= \frac{\Sigma_i \Sigma_j \Sigma_k (Y_{ijk} - \overline{Y}_{ij*})^2}{N - po} \\ &= \frac{(37 - 37.5)^2 + (38 - 37.5)^2 + \cdots + (35 - 35)^2}{53} \\ &= 0.4874. \end{aligned}$$

To obtain the 95% confidence interval for μ_Y we first compute

$$\begin{aligned} K &= S^2_{P*} + S^2_{O*} - S^2_{PO*} \\ &= 371.6 + 17.66 - 2.432 \\ &= 386.8 \end{aligned}$$

and

$$C = \frac{S_{P*}^2\sqrt{F_{0.95:1,9}} + S_{O*}^2\sqrt{F_{0.95:1,2}} - S_{PO*}^2\sqrt{F_{0.95:1,18}}}{K}$$

$$= \frac{371.6\sqrt{5.1174} + 17.66\sqrt{18.513} - 2.432\sqrt{4.4139}}{386.8}$$

$$= 2.356.$$

The 95% confidence interval for μ_Y is

$$L = \overline{Y}^*_{***} - C\sqrt{\frac{K}{por_H}}$$

$$= 35.86 - 2.356\sqrt{\frac{386.8}{10(3)(2.5352)}}$$

$$= 30.5$$

and

$$U = \overline{Y}^*_{***} + C\sqrt{\frac{K}{por_H}}$$

$$= 35.86 + 2.356\sqrt{\frac{386.8}{10(3)(2.5352)}}$$

$$= 41.2.$$

The 95% confidence interval for γ_P requires

$$\widehat{\gamma}_P = \frac{S_{P*}^2 - S_{PO*}^2}{or_H}$$

$$= \frac{371.6 - 2.432}{3(2.5352)}$$

$$= 48.53,$$

$$V_{LP} = G_1^2 S_{P*}^4 + H_3^2 S_{PO*}^4 + G_{13} S_{P*}^2 S_{PO*}^2$$

$$= (0.5269)^2(371.6)^2 + (1.187)^2(2.432)^2 + (-.02358)(371.6)(2.432)$$

$$= 3.83 \times 10^4,$$

and

$$V_{UP} = H_1^2 S_{P*}^4 + G_3^2 S_{PO*}^4 + H_{13} S_{P*}^2 S_{PO*}^2$$

$$= (2.333)^2(371.6)^2 + (0.4290)^2(2.432)^2 + (-0.1800)(371.6)(2.432)$$

$$= 7.51 \times 10^5.$$

The 95% confidence interval for γ_P is

$$L = \widehat{\gamma}_P - \frac{\sqrt{V_{LP}}}{or_H} = 48.53 - \frac{\sqrt{3.83 \times 10^4}}{3(2.5352)} = 22.8$$

and

$$U = \widehat{\gamma}_P + \frac{\sqrt{V_{UP}}}{or_H} = 48.53 + \frac{\sqrt{7.51 \times 10^5}}{3(2.5352)} = 162. \tag{7.16}$$

The 95% confidence interval for γ_M requires

$$\begin{aligned}
\widehat{\gamma}_M &= \frac{S_{O*}^2 + (p-1)S_{PO*}^2 + p(r_H - 1)S_E^2}{pr_H} \\
&= \frac{17.66 + 9(2.432) + 10(1.5352)(0.4874)}{10(2.5352)} \\
&= 1.855, \\
V_{LM} &= G_2^2 S_{O*}^4 + G_3^2(p-1)^2 S_{PO*}^4 + G_4^2 p^2 (r_H - 1)^2 S_E^4 \\
&= (0.7289)^2(17.66)^2 + (0.4290)^2(9)^2(2.432)^2 \\
&\quad + (0.2934)^2(10)^2(1.5352)^2(0.4874)^2 \\
&= 259,
\end{aligned}$$

and

$$\begin{aligned}
V_{UM} &= H_2^2 S_{O*}^4 + H_3^2(p-1)^2 S_{PO*}^4 + H_4^2 p^2 (r_H - 1)^2 S_E^4 \\
&= (38.50)^2(17.66)^2 + (1.187)^2(9)^2(2.432)^2 \\
&\quad + (0.5240)^2(10)^2(1.5352)^2(0.4874)^2 \\
&= 4.63 \times 10^5.
\end{aligned}$$

The 95% confidence interval for γ_M is

$$L = \widehat{\gamma}_M - \frac{\sqrt{V_{LM}}}{pr_H} = 1.855 - \frac{\sqrt{259}}{10(2.5352)} = 1.22$$

and

$$U = \widehat{\gamma}_M + \frac{\sqrt{V_{UM}}}{pr_H}$$
$$= 1.855 + \frac{\sqrt{4.63 \times 10^5}}{10(2.5352)}$$
$$= 28.7. \qquad (7.17)$$

The 95% confidence interval for γ_R requires

$$V_{LR} = po(r_H - 1)S_E^2 + o(1 - G_1)F_3 S_{O*}^2 + o(p-1)S_{PO*}^2$$
$$= 30(1.5352)(0.4874) + 3(0.4731)(39.39)(17.66) + 27(2.432)$$
$$= 1{,}075$$

and

$$V_{UR} = po(r_H - 1)S_E^2 + o(1 + H_1)F_4 S_{O*}^2 + o(p-1)S_{PO*}^2$$
$$= 30(1.5352)(0.4874) + 3(3.333)(0.1750)(17.66) + 27(2.432)$$
$$= 119.$$

The 95% confidence interval for γ_R is

$$L = \frac{p(1 - G_1)(S_{P*}^2 - F_1 S_{PO*}^2)}{V_{LR}}$$
$$= \frac{10(0.4731)[371.6 - 2.929(2.432)]}{1{,}075}$$
$$= 1.60$$

and

$$U = \frac{p(1 + H_1)(S_{P*}^2 - F_2 S_{PO*}^2)}{V_{UR}}$$
$$= \frac{10(3.333)[371.6 - 0.2702(2.432)]}{119}$$
$$= 104. \qquad (7.18)$$

A 95% confidence interval for PTR based on the bounds calculated for γ_M in Equation (7.17) with $k = 6$, $LSL = 18$, and $USL = 58$ is

$$L = \frac{6\sqrt{1.22}}{58 - 18} = 0.166$$

and

$$U = \frac{6\sqrt{28.7}}{58 - 18} = 0.804.$$

Table 7.17. *95% confidence intervals for misclassification rates. (Bounds for δ, δ_c, β, and β_c have been multiplied by 10^6.)*

Parameter	Lower bound	Upper bound
δ	116	39,849
δ_c	116	43,180
δ_{index}	0.105	4.42
β	23	15,007
β_c	74,856	357,792
β_{index}	0.086	0.364

A 95% confidence interval for SNR based on the bounds for γ_R calculated in Equation (7.18) is

$$L = \sqrt{2(1.60)} = 1.79$$

and

$$U = \sqrt{2(104)} = 14.4.$$

The 95% confidence interval for C_p based on the computed interval for γ_P in Equation (7.16) is

$$L = \frac{58 - 18}{6\sqrt{162}} = 0.523$$

and

$$U = \frac{58 - 18}{6\sqrt{22.8}} = 1.40.$$

Table 7.17 reports the 95% confidence intervals for the misclassification rates. These are based on 100,000 simulated GPQ values with $\overline{y}^*_{***} = 35.86$ and $\epsilon = 0.001$. Recall that these GPQs are the USS modification of Table 3.9.

Table 7.18 summarizes the confidence intervals computed in this example. It also reports the generalized confidence intervals based on the USS modification to the GPQs listed in Table 3.8. The R&R graph shown in Figure 7.3 and the interval on δ_{index} in Table 7.17 provide no evidence that the measurement system is capable. The residual plots shown in Figure 7.4 are consistent with the model assumptions.

7.3.4 Modification for Fixed Operators

If operators are fixed, one can apply the USS modification to the formulas reported in Chapter 6. For example, modification of Equation (6.4) yields

$$n^* = \frac{[(o-1) + 2\widehat{\lambda}^*]^2}{(o-1) + 4\widehat{\lambda}^*},$$

Table 7.18. *Comparison of 95% MLS and generalized confidence intervals. (See Section 1.8 for a description of computer programs to perform these computations.)*

Parameter	MLS	GCI
μ_Y	$L = 30.5$	$L = 30.3$
	$U = 41.2$	$U = 41.3$
γ_P	$L = 22.8$	$L = 22.7$
	$U = 162$	$U = 162$
γ_M	$L = 1.22$	$L = 1.16$
	$U = 28.7$	$U = 29.3$
γ_R	$L = 1.60$	$L = 1.66$
	$U = 104$	$U = 85.7$
PTR	$L = 0.166$	$L = 0.162$
	$U = 0.804$	$U = 0.812$
SNR	$L = 1.79$	$L = 1.82$
	$U = 14.4$	$U = 13.1$
C_p	$L = 0.523$	$L = 0.524$
	$U = 1.40$	$U = 1.40$

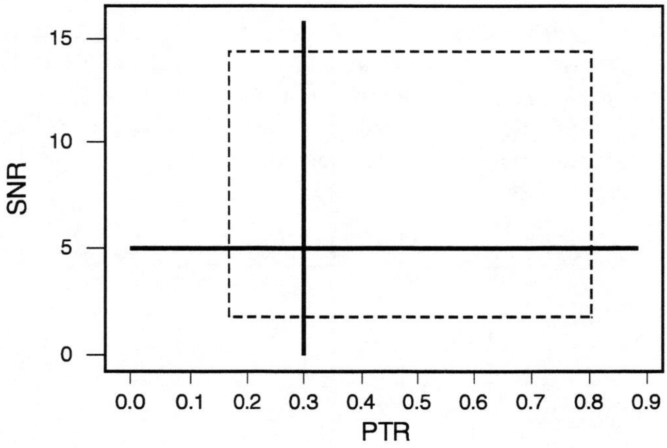

Figure 7.3. *R&R graph for unbalanced two-factor example.*

where

$$\widehat{\lambda}^* = \frac{o-1}{2}\left[\frac{S^2_{O*}}{S^2_{PO*}}\left\{\frac{(p-1)(o-1)-2}{(p-1)(o-1)}\right\} - 1\right] \tag{7.19}$$

for $(p-1)(o-1) > 2$.

Recall that in Chapter 6 $U = (o-1)S^2_O/\theta_{PO}$ has an exact noncentral chi-squared distribution. The definition of n^* in Chapter 6 was determined by matching moments of U

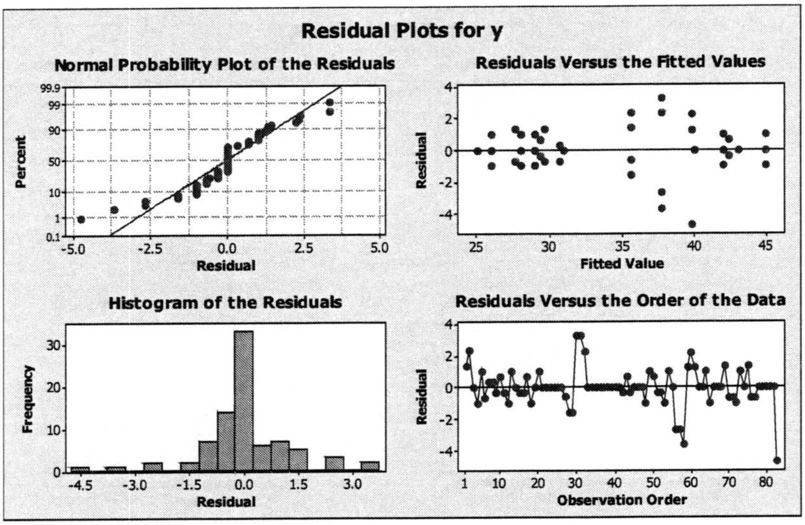

Figure 7.4. *Residual plots for the two-factor unbalanced example.*

with a central chi-squared distribution. In the unbalanced case, the random variable $W = (o-1)S^2_{O*}/\theta_{PO*}$ is not an exact noncentral chi-squared variable. Thus, the approximation in Equation (7.19) does not match the first two moments of W exactly. However, this approximation should work well if the data are not too unbalanced. Gong, Burdick, and Quiroz [22] provided simulations that demonstrate the performance of this approximation.

For the numerical example in Section 7.3.3,

$$\begin{aligned}\widehat{\lambda}^* &= \frac{o-1}{2}\left[\frac{S^2_{O*}}{S^2_{PO*}}\left\{\frac{(p-1)(o-1)-2}{(p-1)(o-1)}\right\}-1\right]\\ &= \frac{2}{2}\left[\frac{17.66}{2.432}\left\{\frac{18-2}{18}\right\}-1\right]\\ &= 5.45\end{aligned}$$

and

$$\begin{aligned}n^* &= \frac{[(o-1)+2\widehat{\lambda}^*]^2}{(o-1)+4\widehat{\lambda}^*}\\ &= \frac{[2+2(5.45)]^2}{2+4(5.45)}\\ &= 7.0.\end{aligned}$$

We now use $n^* = 7$ instead of $o-1=2$ for determining the constants G_2, H_2, F_3, and F_4 in Table 7.14 when operators are fixed. The USS modification can also be applied to the GPQs in Tables 6.10 and 6.11.

Table 7.19. *USS ANOVA for model* (7.20).

Source of variation	Degrees of freedom	Mean square	Expected mean square
Parts (P)	$p-1$	S_{P*}^2	$\theta_{P*} = \sigma_E^2 + or_H\sigma_P^2$
Operators (O)	$o-1$	S_{O*}^2	$\theta_{O*} = \sigma_E^2 + pr_H\sigma_O^2$
Replicates (E)	$N-p-o+1$	S_{E*}^2	$\theta_{E*} = \sigma_E^2$

7.4 Unbalanced Two-Factor Random Model with No Interaction

In Chapter 5 we presented confidence intervals for the balanced two-factor random model with no interaction. The unbalanced two-factor random model with no interaction is

$$Y_{ijk} = \mu_Y + P_i + O_j + E_{ijk}, \qquad (7.20)$$
$$i = 1, \ldots, p, \quad j = 1, \ldots, o, \quad k = 1, \ldots, r_{ij} > 0,$$

where μ_Y is a constant and P_i, O_j, and E_{ijk} are jointly independent normal random variables with means of zero and variances σ_P^2, σ_O^2, and σ_E^2, respectively.

The approach for analyzing an unbalanced design is to use the balanced model equations with the USS modification. The estimator for σ_E^2 is formed by pooling S_{PO*}^2 and S_E^2 from the unbalanced two-factor model with interaction. In particular, the estimator for σ_E^2 in model (7.20) is

$$S_{E*}^2 = \frac{(p-1)(o-1)S_{PO*}^2 + (N-po)S_E^2}{N-p-o+1}, \qquad (7.21)$$

where $N = \Sigma_i \Sigma_j r_{ij}$, and S_{PO*}^2 and S_E^2 are defined in Table 7.11.

Tables 7.19 and 7.20 report the USS ANOVA and mean square definitions for model (7.20). The definitions for S_{P*}^2 and S_{O*}^2 are the same as those given in Table 7.11. All cells must have at least one observation in order to compute the USS mean squares. Under the assumptions of model (7.20), $(p-1)S_{P*}^2/\theta_{P*}$, $(o-1)S_{O*}^2/\theta_{O*}$, and $(N-p-o+1)S_{E*}^2/\theta_{E*}$ have approximate chi-squared distributions with $(p-1)$, $(o-1)$, and $(N-p-o+1)$ degrees of freedom, respectively. These random variables are not generally independent.

7.4.1 Confidence Intervals for Gauge R&R Parameters

Table 7.21 reports confidence intervals for the gauge R&R parameters in model (7.20). They are formed from the intervals in Chapter 5 using the USS modification where we replace r, S_P^2, S_O^2, and S_E^2 with r_H, S_{P*}^2, S_{O*}^2, and S_{E*}^2, respectively. Constants and definitions used in the table are reported in Tables 7.22 and 7.23. Gong, Burdick, and Quiroz [22] report computer simulations that demonstrate the USS confidence intervals generally work well in typical gauge R&R studies.

Table 7.20. *USS mean squares and means for model* (7.20).

MS	Definition
S^2_{P*}	$\dfrac{or_H \Sigma_i (\overline{Y}^*_{i**} - \overline{Y}^*_{***})^2}{p-1}$
S^2_{O*}	$\dfrac{pr_H \Sigma_j (\overline{Y}^*_{*j*} - \overline{Y}^*_{***})^2}{o-1}$
S^2_{E*}	$\dfrac{(p-1)(o-1)S^2_{PO*} + (N-po)S^2_E}{N-p-o+1}$
S^2_{PO*}	$\dfrac{r_H \Sigma_i \Sigma_j (\overline{Y}_{ij*} - \overline{Y}^*_{i**} - \overline{Y}^*_{*j*} + \overline{Y}^*_{***})^2}{(p-1)(o-1)}$
S^2_E	$\dfrac{\Sigma_i \Sigma_j \Sigma_k (Y_{ijk} - \overline{Y}_{ij*})^2}{N-po}$
$\overline{Y}_{ij*}$	$\dfrac{\Sigma_k Y_{ijk}}{r_{ij}}$
$\overline{Y}^*_{i**}$	$\dfrac{\Sigma_j \overline{Y}_{ij*}}{o}$
$\overline{Y}^*_{*j*}$	$\dfrac{\Sigma_i \overline{Y}_{ij*}}{p}$
$\overline{Y}^*_{***}$	$\dfrac{\Sigma_i \Sigma_j \overline{Y}_{ij*}}{po}$
r_H	$\dfrac{po}{\Sigma_i \Sigma_j 1/r_{ij}}$
N	$\Sigma_i \Sigma_j r_{ij}$

Table 7.21. *Confidence intervals for gauge R&R parameters in model* (7.20).

Parameter	Balanced equation	USS modification Lower bound	USS modification Upper bound
μ_Y	(5.2)	$\overline{Y}^*_{***} - C\sqrt{\frac{K}{por_H}}$	$\overline{Y}^*_{***} + C\sqrt{\frac{K}{por_H}}$
γ_P	(5.3)	$\widehat{\gamma}_P - \dfrac{\sqrt{V_{LP}}}{or_H}$	$\widehat{\gamma}_P + \dfrac{\sqrt{V_{UP}}}{or_H}$
$\gamma_M = \sigma^2_O + \sigma^2_E$	(5.4)	$\widehat{\gamma}_M - \dfrac{\sqrt{V_{LM}}}{pr_H}$	$\widehat{\gamma}_M + \dfrac{\sqrt{V_{UM}}}{pr_H}$
γ_R	(5.5)	$\dfrac{V_{LR1}}{V_{LR2}}$	$\dfrac{V_{UR1}}{V_{UR2}}$

Table 7.22. *Terms used in Table* 7.21.

Term	Definition
K	$S^2_{P*} + S^2_{O*} - S^2_{E*}$
C	$\dfrac{S^2_{P*}\sqrt{F_{1-\alpha:1,p-1}} + S^2_{O*}\sqrt{F_{1-\alpha:1,o-1}} - S^2_{E*}\sqrt{F_{1-\alpha:1,N-p-o+1}}}{K}$
$\widehat{\gamma}_P$	$\dfrac{S^2_{P*} - S^2_{E*}}{or_H}$
V_{LP}	$G_1^2 S^4_{P*} + H_3^2 S^4_{E*} + G_{13} S^2_{P*} S^2_{E*}$
V_{UP}	$H_1^2 S^4_{P*} + G_3^2 S^4_{E*} + H_{13} S^2_{P*} S^2_{E*}$
$\widehat{\gamma}_M$	$\dfrac{S^2_{O*} + (pr_H - 1) S^2_{E*}}{pr_H}$
V_{LM}	$G_2^2 S^4_{O*} + G_3^2 (pr_H - 1)^2 S^4_{E*}$
V_{UM}	$H_2^2 S^4_{O*} + H_3^2 (pr_H - 1)^2 S^4_{E*}$
V_{LR1}	$p(1 - G_1) S^4_{P*} - p S^2_{P*} S^2_{E*} + p[F_1 - (1 - G_1) F_1^2] S^4_{E*}$
V_{LR2}	$o(pr_H - 1) S^2_{P*} S^2_{E*} + o(1 - G_1) F_3 S^2_{P*} S^2_{O*}$
V_{UR1}	$p(1 + H_1) S^4_{P*} - p S^2_{P*} S^2_{E*} + p[F_2 - (1 + H_1) F_2^2] S^4_{E*}$
V_{UR2}	$o(pr_H - 1) S^2_{P*} S^2_{E*} + o(1 + H_1) F_4 S^2_{P*} S^2_{O*}$

7.4.2 Intervals for Misclassification Rates

The GPQs for computing misclassification rates are obtained by applying the USS modification to the GPQs shown in Table 5.7.

7.4.3 Numerical Example

The computed confidence intervals for the data in Table 7.15 using model (7.20) are shown in Tables 7.24 and 7.25. The generalized confidence intervals are based on 100,000 simulated GPQ values with $\epsilon = 0.001$. To compute these intervals, we first obtain

$$\begin{aligned} S^2_{E*} &= \frac{(p-1)(o-1)S^2_{PO*} + (N - po)S^2_E}{N - p - o + 1} \\ &= \frac{18(2.432) + 53(0.4874)}{83 - 10 - 3 + 1} \\ &= 0.980. \end{aligned}$$

Table 7.23. *Constants used in Table 7.21. Values are for* $\alpha = 0.05$, $p = 10$, $o = 3$, *and* $N = 83$.

Constant	Definition	Value
G_1	$1 - F_{\alpha/2:\infty,p-1}$	0.5269
G_2	$1 - F_{\alpha/2:\infty,o-1}$	0.7289
G_3	$1 - F_{\alpha/2:\infty,N-p-o+1}$	0.2619
H_1	$F_{1-\alpha/2:\infty,p-1} - 1$	2.333
H_2	$F_{1-\alpha/2:\infty,o-1} - 1$	38.50
H_3	$F_{1-\alpha/2:\infty,N-p-o+1} - 1$	0.4317
F_1	$F_{1-\alpha/2:p-1,N-p-o+1}$	2.299
F_2	$F_{\alpha/2:p-1,N-p-o+1}$	0.2914
F_3	$F_{1-\alpha/2:p-1,o-1}$	39.39
F_4	$F_{\alpha/2:p-1,o-1}$	0.1750
G_{13}	$\dfrac{(F_1 - 1)^2 - G_1^2 F_1^2 - H_3^2}{F_1}$	0.01469
H_{13}	$\dfrac{(1 - F_2)^2 - H_1^2 F_2^2 - G_3^2}{F_2}$	−0.09818

Table 7.24. *Comparison of MLS and generalized confidence intervals. (See Section* 1.8 *for a description of computer programs to perform these computations.)*

Parameter	MLS	GCI
μ_Y	$L = 30.5$	$L = 30.4$
	$U = 41.2$	$U = 41.4$
γ_P	$L = 23.0$	$L = 23.0$
	$U = 163$	$U = 163$
γ_M	$L = 1.07$	$L = 1.06$
	$U = 28.5$	$U = 28.6$
γ_R	$L = 1.65$	$L = 1.73$
	$U = 121$	$U = 96.7$
PTR	$L = 0.155$	$L = 0.154$
	$U = 0.800$	$U = 0.802$
SNR	$L = 1.82$	$L = 1.86$
	$U = 15.5$	$U = 13.9$
C_p	$L = 0.523$	$L = 0.523$
	$U = 1.39$	$U = 1.39$

7.4.4 Modification for Fixed Operators

The modification for fixed operators is handled as described in Section 7.3.4 with the obvious substitutions. This gives

$$n^* = \frac{[(o-1) + 2\widehat{\lambda}^*]^2}{(o-1) + 4\widehat{\lambda}^*},$$

Table 7.25. *95% confidence intervals for misclassification rates. (Bounds for δ, δ_c, β, and β_c have been multiplied by 10^6.)*

Parameter	Lower bound	Upper bound
δ	107	39,390
δ_c	107	42,795
δ_{index}	0.096	4.09
β	25	14,700
β_c	69,992	355,251
β_{index}	0.080	0.362

where

$$\widehat{\lambda}^* = \frac{o-1}{2}\left[\frac{S_{O*}^2}{S_{E*}^2}\left\{\frac{(N-p-o+1)-2}{N-p-o+1}\right\}-1\right].$$

This approximation does not match the first two moments of the chi-squared distribution exactly, but it should provide good results in most practical situations. Gong, Burdick, and Quiroz [22] provide simulations of this method in several typical gauge R&R designs. The USS modification can also be applied to the GPQs shown in Tables 6.19 and 6.20.

7.5 Summary

In this chapter we presented the USS modification for analyzing unbalanced ANOVA models. The method is very general and can be applied to any design for which balanced model equations are available. Examples were provided for both one-factor and two-factor designs. In the next chapter, we provide a general strategy for constructing confidence intervals in any ANOVA model. This includes both random and mixed models with either balanced or unbalanced data sets.

7.6 Appendix: Generalized Confidence Interval for γ_P in Model (7.1)

In this appendix, we describe the approach recommended by Park and Burdick [56] for constructing a GCI for $\gamma_P = \sigma_P^2$ in the unbalanced one-factor random design. These results can be applied to any mixed model with two error terms.

To begin, we write the unbalanced one-factor random model in model (7.1) in matrix notation as

$$\boldsymbol{Y} = \boldsymbol{\mu}_Y + \boldsymbol{B}\boldsymbol{P} + \boldsymbol{E},$$

where $\mathbf{Y}$ is an $N \times 1$ vector of observations, $N = \Sigma_i r_i$, $\boldsymbol{\mu}_Y$ is an $N \times 1$ vector with the scalar μ_Y in every position, $\boldsymbol{B} = \oplus_{i=1}^{p} \mathbf{1}_{r_i}$ is an $N \times p$ design matrix, where $\mathbf{1}_{r_i}$ is a $r_i \times 1$ column vector of ones, $\boldsymbol{P}$ is a $p \times 1$ vector with elements $P_1, \ldots, P_p$, and $\boldsymbol{E}$ is an $N \times 1$ vector of random error terms. Under the assumptions of the model, $\boldsymbol{Y}$ has a multivariate normal distribution with mean $\boldsymbol{\mu}_Y$ and covariance matrix $\gamma_P \boldsymbol{B}\boldsymbol{B}' + \sigma_E^2 \boldsymbol{I}_N$, where $\boldsymbol{I}_N$ is an $N \times N$ identity matrix.

Park and Burdick [56] proposed a GPQ for constructing a generalized confidence interval on γ_P using results provided by Olsen, Seely, and Birkes [54]. Let $\boldsymbol{H} = \boldsymbol{F}\boldsymbol{B}\boldsymbol{B}'\boldsymbol{F}$, where $\boldsymbol{F} = \boldsymbol{X}(\boldsymbol{X}'\boldsymbol{X})^+\boldsymbol{X}' - \boldsymbol{1}_N(\boldsymbol{1}_N'\boldsymbol{1}_N)^+\boldsymbol{1}_N'$, $\boldsymbol{X} = (\boldsymbol{1}_N, \boldsymbol{B}\boldsymbol{B}')$ is the horizontal concatenation of matrices $\boldsymbol{1}_N$ and $\boldsymbol{B}\boldsymbol{B}'$, and $^+$ denotes a Moore–Penrose inverse. Let $d_1, d_2, \ldots, d_m$ denote the distinct positive eigenvalues of $\boldsymbol{H}$ and let n_l be the multiplicity of d_l for $l = 1, 2, \ldots, m$. Note that rank$(\boldsymbol{H}) = \sum_l n_l = p-1$. Consistent with LaMotte [37], Olsen, Seely, and Birkes show that $Q_l/(\sigma_E^2 + d_l\gamma_P)$ are chi-squared random variables with degrees of freedom n_l for $l = 1, \ldots, m$, where $Q_l = \boldsymbol{Y}'\boldsymbol{F}'\boldsymbol{E}_l\boldsymbol{F}\boldsymbol{Y}$ and $\boldsymbol{E}_l$ is the orthogonal projection operator of the eigenspace of d_l. Finally, using this notation $S_E^2 = \boldsymbol{Y}'[\boldsymbol{I}_N - \boldsymbol{X}(\boldsymbol{X}'\boldsymbol{X})^+\boldsymbol{X}']\boldsymbol{Y}/(N-p)$. The random variables $S_E^2, Q_1, \ldots, Q_m$ are mutually independent and so it follows that

$$\sum_{l=1}^{m} \frac{Q_l}{\sigma_E^2 + d_l\gamma_P} = W_1$$

has a chi-squared distribution with $p-1$ degrees of freedom.

To construct a generalized confidence interval for γ_P, define T as the solution for γ_P in the nonlinear equation

$$\sum_{i=1}^{m} \frac{q_l}{(N-p)s_E^2/W_2 + d_l\gamma_P} = W_1, \tag{7.22}$$

where q_l and s_E^2 are observed values of Q_l and S_E^2, respectively, and W_1 and W_2 are jointly independent chi-squared variables with degrees of freedom $p-1$ and $N-p$, respectively. Note that the distribution of T is completely determined by the joint distribution of W_1 and W_2 and is free of the parameters contained in the model. We now compute 100,000 values of T by simulating 100,000 values each of W_1 and W_2. If $W_1 > W_2\sum_l q_l/[(N-p)s_E^2]$, then set $T = 0$ since $\gamma_P \geq 0$. If $W_1 \leq W_2\sum_l q_l/[(N-p)s_E^2]$, then the bisection method is used to solve the nonlinear equation in (7.22). The resulting 100,000 values of T are ordered from least to greatest. The approximate $100(1-\alpha)\%$ confidence interval for γ_P is from $L = T_{\alpha/2}$ to $U = T_{1-\alpha/2}$, where $T_{\alpha/2}$ is the value of T in position $100{,}000 \times \alpha/2$ of the ordered set and $T_{1-\alpha/2}$ is the value of T in position $100{,}000 \times (1-\alpha/2)$.

Simulation results in Burdick and Park [56] show that the generalized interval for γ_P is comparable to Equation (7.4) unless γ_R is small. In this case, computation of the GCI is recommended because Equation (7.4) will have a confidence level of less than the stated level.

Chapter 8

Strategies for Constructing Intervals with ANOVA Models

8.1 Introduction

Although the two-factor model is the most common gauge R&R design, more complex ANOVA designs are needed to properly model some measurement systems. In this chapter we present general strategies for constructing confidence intervals in any ANOVA design. These strategies have been used throughout the book, and we now describe them in more general notation. We begin with balanced random models, and then we describe adjustments for unbalanced and mixed models. We demonstrate how to apply these strategies with three different ANOVA designs.

8.2 General Strategies

Consider a balanced random model with Q sources of variation represented by the ANOVA shown in Table 8.1. Assuming all random errors are independent and normally distributed, the $n_q S_q^2/\theta_q$ are jointly independent and each $n_q S_q^2/\theta_q$ has a chi-squared distribution with n_q degrees of freedom for $q = 1, \ldots, Q$ (see, e.g., Hocking [31, p. 460]). These results are

Table 8.1. *General ANOVA for balanced random model.*

Source of variation	Degrees of freedom	Mean square	Expected mean square
Factor 1	n_1	S_1^2	θ_1
Factor 2	n_2	S_2^2	θ_2
$\vdots$	$\vdots$	$\vdots$	$\vdots$
Replicates	n_Q	S_Q^2	θ_Q

Table 8.2. *Parameter functions of* θ_q.

Parameter	Function
γ_1	$\sum_{q=1}^{Q} c_q\theta_q$
γ_2	$\sum_{r=1}^{P} d_r\theta_r - \sum_{t=P+1}^{Q} e_t\theta_t$
γ_3	$\dfrac{\sum_{r=1}^{P} d_r\theta_r}{\sum_{t=P+1}^{Q} e_t\theta_t}$
γ_4	$\dfrac{\sum_{r=1}^{P} d_r\theta_r - \sum_{t=P+1}^{Q} e_t\theta_t}{\sum_{q=1}^{Q} c_q\theta_q}$

used to construct confidence intervals for all four parameters shown in Table 8.2, where c_q, d_r, and $e_t \geq 0$. Note that γ_3 and γ_4 differ in two respects. First, γ_3 has nonnegative linear combinations in both the numerator and the denominator, whereas γ_4 allows negative linear combinations in the numerator. Second, the same expected mean square cannot appear in both the numerator and the denominator of γ_3, but it can in γ_4. Although γ_3 is a special case of γ_4, it is useful to consider the two cases separately. Note that neither γ_3 nor γ_4 allows negative coefficients in the denominator.

All the gauge R&R parameters can be represented as one of the functions shown in Table 8.2. To demonstrate, consider the random two-factor model with interaction presented in Chapter 3. Table 8.3 reports the ANOVA for this model. We have numbered the mean squares (S_q^2) and the expected mean squares (θ_q) sequentially to correspond to the general notation in Table 8.1.

The parameter $\gamma_P = \sigma_P^2$ is written in terms of the expected mean squares as

$$\gamma_P = \frac{\theta_1 - \theta_3}{or}. \tag{8.1}$$

Note that Equation (8.1) can be represented as γ_2 in Table 8.2 with $Q = 4$, $P = 2$, $d_1 = 1/(or)$, $d_2 = 0$, $e_3 = 1/(or)$, and $e_4 = 0$. The parameter $\gamma_M = \sigma_O^2 + \sigma_{PO}^2 + \sigma_E^2$ is written in terms of mean squares as

$$\gamma_M = \frac{\theta_2 + (p-1)\theta_3 + p(r-1)\theta_4}{pr}. \tag{8.2}$$

Thus, γ_M can be represented as γ_1 with $c_1 = 0$, $c_2 = 1/(pr)$, $c_3 = (p-1)/(pr)$, and

Table 8.3. *ANOVA for random two-factor model with interaction.*

Source of variation	Degrees of freedom	Mean square	Expected mean square
Parts (P)	$n_1 = p - 1$	S_1^2	$\theta_1 = \sigma_E^2 + r\sigma_{PO}^2 + or\sigma_P^2$
Operators (O)	$n_2 = o - 1$	S_2^2	$\theta_2 = \sigma_E^2 + r\sigma_{PO}^2 + pr\sigma_O^2$
P×O	$n_3 = (p-1)(o-1)$	S_3^2	$\theta_3 = \sigma_E^2 + r\sigma_{PO}^2$
Replicates	$n_4 = po(r-1)$	S_4^2	$\theta_4 = \sigma_E^2$

Table 8.4. *GPQs for parameter functions of* θ_q.

Parameter	GPQ
γ_1	$\sum_{q=1}^{Q} \frac{c_q n_q s_q^2}{W_q}$
γ_2	$\sum_{r=1}^{P} \frac{d_r n_r s_r^2}{W_r} - \sum_{t=P+1}^{Q} \frac{e_t n_t s_t^2}{W_t}$
γ_3	$\frac{\sum_{r=1}^{P} d_r n_r s_r^2 / W_r}{\sum_{t=P+1}^{Q} e_t n_t s_t^2 / W_t}$
γ_4	$\frac{\sum_{r=1}^{P} d_r n_r s_r^2 / W_r - \sum_{t=P+1}^{Q} e_t n_t s_t^2 / W_t}{\sum_{q=1}^{Q} c_q n_q s_q^2 / W_q}$

$c_4 = (r-1)/r$. The parameter $\gamma_R = \gamma_P/\gamma_M$ can be represented as γ_4 using the definitions in Equations (8.1) and (8.2).

GCIs can be constructed for all the parameters in Table 8.2. The GPQs for these parameters are shown in Table 8.4, where s_q^2 is the realized value of S_q^2 and W_1, W_2, $\ldots$, W_Q are independent chi-squared random variables with n_q degrees of freedom for $q = 1, \ldots, Q$. These GPQs also correspond to the surrogate variables recommended by Chiang [14]. Alternatively, MLS methods can be applied if one desires closed-form intervals. We now describe these MLS methods.

8.2.1 MLS Intervals for γ_1

The MLS procedure used to construct confidence intervals for $\gamma_1 = \sum_{q=1}^{Q} c_q \theta_q$ where $c_q \geq 0$ is the method of Graybill and Wang [25]. We have used this method throughout the book to construct intervals for γ_M. The $100(1-\alpha)\%$ Graybill–Wang interval for γ_1 is

$$L = \widehat{\gamma}_1 - \sqrt{\sum_{q=1}^{Q} G_q^2 c_q^2 S_q^4}$$

and

$$U = \widehat{\gamma}_1 + \sqrt{\sum_{q=1}^{Q} H_q^2 c_q^2 S_q^4}, \tag{8.3}$$

where

$$\widehat{\gamma}_1 = \sum_{q=1}^{Q} c_q S_q^2,$$
$$G_q = 1 - F_{\alpha/2:\infty,n_q},$$

and

$$H_q = F_{1-\alpha/2:\infty,n_q} - 1.$$

The length of the interval in Equation (8.3) and the length of the GCI for γ_1 are greatly affected by the minimum value of the n_q. In designing an experiment to estimate γ_1, the minimum value of n_q should exceed four to provide intervals that are short enough to be useful.

Another closed-form interval for γ_1 can be constructed using the method proposed by Satterthwaite [57, 58]. This interval is formed by matching the first two moments of the statistic $m\widehat{\gamma}_1/\gamma_1$ with those of a chi-squared random variable with m degrees of freedom. An approximate $100(1-\alpha)\%$ confidence interval for γ_1 based on this approximation is

$$L = (1 - G_m)\widehat{\gamma}_1$$

and

$$U = (1 + H_m)\widehat{\gamma}_1, \tag{8.4}$$

where

$$G_m = 1 - F_{\alpha/2:\infty,m},$$
$$H_m = F_{1-\alpha/2:\infty,m} - 1,$$

and

$$m = \frac{\widehat{\gamma}_1^2}{\sum_{q=1}^{Q} c_q^2 S_q^4 / n_q}.$$

In practice, one should truncate m to the greatest integer less than or equal to m. Equation (8.4) provides an interval that generally is close to the stated confidence level when the n_q are of similar magnitude or when all n_q are large. In cases in which these conditions are not met, Equation (8.4) can yield confidence coefficients much less than the stated level. In such a case, it is recommended to use either Equation (8.3) or the GCI. The Satterthwaite interval can be recommended if the n_q are all small (say, three or less). In this case, the Satterthwaite interval provides a confidence coefficient close to the stated level and generally yields a shorter interval than the other two methods.

Finally, if γ_1 has only one c_q value greater than zero, an exact $100(1-\alpha)\%$ interval on $c_q\theta_q$ is

$$L = (1 - G_q)c_q S_q^2$$

and

$$U = (1 + H_q)c_q S_q^2, \tag{8.5}$$

where G_q and H_q are defined in Equation (8.3).

To demonstrate these general results, we will derive the confidence interval for γ_M shown in Equation (3.4). As shown in Equation (8.2), $\gamma_M = \gamma_1$, where $Q = 4$, $c_1 = 0$, $c_2 = 1/(pr)$, $c_3 = (p-1)/(pr)$, and $c_4 = (r-1)/r$. Thus,

$$\begin{aligned}
\widehat{\gamma}_M &= \widehat{\gamma}_1 \\
&= \frac{S_2^2}{pr} + \frac{(p-1)S_3^2}{pr} + \frac{(r-1)S_4^2}{r} \\
&= \frac{S_2^2 + (p-1)S_3^2 + p(r-1)S_E^2}{pr}.
\end{aligned}$$

This is the point estimator for γ_M shown in Table 3.6 with $S_2^2 = S_O^2$, $S_3^2 = S_{PO}^2$, and $S_4^2 = S_E^2$. The lower bound for γ_M is obtained from Equation (8.3) as

$$\begin{aligned}
L &= \widehat{\gamma}_M - \sqrt{\frac{G_2^2 S_2^4}{p^2r^2} + \frac{G_3^2(p-1)^2 S_3^4}{p^2r^2} + \frac{G_4^2 p^2(r-1)^2 S_4^4}{p^2r^2}} \\
&= \widehat{\gamma}_M - \frac{\sqrt{V_{LM}}}{pr},
\end{aligned}$$

where V_{LM} is defined in Equation (3.4). In a similar manner, the upper bound for γ_1 reduces to the upper bound for γ_M shown in Equation (3.4).

8.2.2 MLS Intervals for γ_2

We first consider the special case of $\gamma_2 = \sum_{r=1}^{P} d_r\theta_r - \sum_{t=P+1}^{Q} e_t\theta_t$, where $Q = 2$ and $P = 1$. Here $\gamma_2 = d_1\theta_1 - e_2\theta_2$. As discussed in Section 2.3.2, several methods have been proposed for constructing confidence intervals on a difference of two expected mean squares. In previous chapters we used the MLS interval proposed by Ting et al. [64] and the method developed independently by Tukey [66] and Williams [73] (see, e.g., Equations (2.3) and (2.4)). The $100(1-\alpha)\%$ MLS interval for $d_1\theta_1 - e_2\theta_2$ is

$$L = \widehat{\gamma}_2 - \sqrt{V_L}$$

and

$$U = \widehat{\gamma}_2 + \sqrt{V_U}, \tag{8.6}$$

where

$$
\begin{aligned}
\widehat{\gamma}_2 &= d_1 S_1^2 - e_2 S_2^2,\\
V_L &= G_1^2 d_1^2 S_1^4 + H_2^2 e_2^2 S_2^4 + G_{12} d_1 e_2 S_1^2 S_2^2,\\
V_U &= H_1^2 d_1^2 S_1^4 + G_2^2 e_2^2 S_2^4 + H_{12} d_1 e_2 S_1^2 S_2^2,\\
G_1 &= 1 - F_{\alpha/2:\infty,n_1},\\
G_2 &= 1 - F_{\alpha/2:\infty,n_2},\\
H_1 &= F_{1-\alpha/2:\infty,n_1} - 1,\\
H_2 &= F_{1-\alpha/2:\infty,n_2} - 1,\\
G_{12} &= \frac{(F_{1-\alpha/2:n_1,n_2} - 1)^2 - G_1^2 F_{1-\alpha/2:n_1,n_2}^2 - H_2^2}{F_{1-\alpha/2:n_1,n_2}},
\end{aligned}
$$

and

$$
H_{12} = \frac{(1 - F_{\alpha/2:n_1,n_2})^2 - H_1^2 F_{\alpha/2:n_1,n_2}^2 - G_2^2}{F_{\alpha/2:n_1,n_2}}.
$$

To demonstrate, these results can be applied to construct the confidence interval for γ_P shown in Equation (3.3). Here $\gamma_P = \gamma_2$, where $d_1 = e_2 = 1/(or)$, $\theta_1 = \theta_P$, and $\theta_2 = \theta_{PO}$. Making these substitutions and replacing S_1^2 with S_P^2 and S_2^2 with S_{PO}^2 in Equation (8.6) provides the interval for γ_P shown in Equation (3.3).

The $100(1-\alpha)\%$ Tukey–Williams interval for $d_1\theta_1 - e_2\theta_2$ is

$$
L = (d_1 S_1^2 - e_2 S_2^2 F_{1-\alpha/2:n_1,n_2})(1 - G_1)
$$

and

$$
U = (d_1 S_1^2 - e_2 S_2^2 F_{\alpha/2:n_1,n_2})(1 + H_1), \tag{8.7}
$$

where G_1 and H_1 are defined in Equation (8.6).

The MLS method can be extended for any value of Q or P, but the algebra gets more complex. The $100(1-\alpha)\%$ MLS interval for γ_2 with $Q > 2$ is

$$
L = \widehat{\gamma}_2 - \sqrt{V_L}
$$

and

$$
U = \widehat{\gamma}_2 + \sqrt{V_U}, \tag{8.8}
$$

where

$$
\begin{aligned}
\widehat{\gamma}_2 &= \sum_{r=1}^{P} d_r S_r^2 - \sum_{t=P+1}^{Q} e_t S_t^2,\\
V_L &= \sum_{r=1}^{P} G_r^2 d_r^2 S_r^4 + \sum_{t=P+1}^{Q} H_t^2 e_t^2 S_t^4 + \sum_{r=1}^{P} \sum_{t=P+1}^{Q} G_{rt} d_r e_t S_r^2 S_t^2\\
&\quad + \sum_{r=1}^{P-1} \sum_{u>r}^{P} G_{ru}^* d_r d_u S_r^2 S_u^2,
\end{aligned}
$$

$$V_U = \sum_{r=1}^{P} H_r^2 d_r^2 S_r^4 + \sum_{t=P+1}^{Q} G_t^2 e_t^2 S_t^4 + \sum_{r=1}^{P} \sum_{t=P+1}^{Q} H_{rt} d_r e_t S_r^2 S_t^2$$
$$+ \sum_{t=P+1}^{Q-1} \sum_{v>t}^{Q} H_{tv}^* e_t e_v S_t^2 S_v^2,$$

$$G_r = 1 - F_{\alpha/2:\infty,n_r} \quad (r = 1, \ldots, P),$$
$$H_t = F_{1-\alpha/2:\infty,n_t} - 1 \quad (t = P+1, \ldots, Q),$$
$$G_{rt} = \frac{(F_{1-\alpha/2:n_r,n_t} - 1)^2 - G_r^2 F_{1-\alpha/2:n_r,n_t}^2 - H_t^2}{F_{1-\alpha/2:n_r,n_t}},$$
$$G_{ru}^* = \left[\left(1 - F_{\alpha/2:\infty,n_r+n_u}\right)^2 \frac{(n_r+n_u)^2}{n_r n_u} - \frac{G_r^2 n_r}{n_u} - \frac{G_u^2 n_u}{n_r}\right] / (P-1)$$
$$(u = r+1, \ldots, P),$$
$$H_r = F_{1-\alpha/2:\infty,n_r} - 1 \quad (r = 1, \ldots, P),$$
$$G_t = 1 - F_{\alpha/2:\infty,n_t} \quad (t = P+1, \ldots, Q),$$
$$H_{rt} = \frac{(1 - F_{\alpha/2:n_r,n_t})^2 - H_r^2 F_{\alpha/2:n_r,n_t}^2 - G_t^2}{F_{\alpha/2:n_r,n_t}},$$

and

$$H_{tv}^* = \left[\left(1 - F_{\alpha/2:\infty,n_t+n_v}\right)^2 \frac{(n_t+n_v)^2}{n_t n_v} - \frac{G_t^2 n_t}{n_v} - \frac{G_v^2 n_v}{n_t}\right] / (Q-P-1)$$
$$(v = t+1, \ldots, Q).$$

When $Q = 2$ and $P = 1$, Equation (8.8) is equivalent to Equation (8.6). Ting et al. [64] used computer simulation to demonstrate that Equation (8.8) provides confidence coefficients close to the stated level over a wide range of conditions. Intervals computed with Equation (8.8) are comparable to the GCIs based on the pivotal quantities shown in Table 8.4.

The lengths of the MLS intervals and GCIs for γ_2 depend on the minimum value of the n_q. Seely and Lee [60] proposed an approach that will provide shorter intervals than both of these methods when all the n_q are small. This approach first combines expected mean squares of like sign using the Satterthwaite approximation. In particular, let

$$\nu_1 = \sum_{r=1}^{P} d_r \theta_r$$

and

$$\nu_2 = \sum_{t=P+1}^{Q} e_t \theta_t \tag{8.9}$$

so that $\gamma_2 = \nu_1 - \nu_2$. The estimators

$$\widehat{\nu}_1 = \sum_{r=1}^{P} d_r S_r^2$$

and

$$\widehat{\nu_2} = \sum_{t=P+1}^{Q} e_t S_t^2 \tag{8.10}$$

are now approximated as chi-squared variables using the Satterthwaite approximation. That is, $m_1\widehat{\nu_1}/\nu_1$ and $m_2\widehat{\nu_2}/\nu_2$ are approximated as chi-squared random variables with m_1 and m_2 degrees of freedom, respectively, where

$$m_1 = \frac{\widehat{\nu_1^2}}{\sum_{r=1}^{P} d_r^2 S_r^4/n_r}$$

and

$$m_2 = \frac{\widehat{\nu_2^2}}{\sum_{t=P+1}^{Q} e_t^2 S_t^4/n_t}. \tag{8.11}$$

Seely and Lee then use Equation (8.7) to provide the approximate $100(1-\alpha)\%$ interval for $\gamma_2 = \nu_1 - \nu_2$,

$$L = (\widehat{\nu_1} - \widehat{\nu_2} F_{1-\alpha/2:m_1,m_2}) F_{\alpha/2:\infty,m_1}$$

and

$$U = (\widehat{\nu_1} - \widehat{\nu_2} F_{\alpha/2:m_1,m_2}) F_{1-\alpha/2:\infty,m_1}, \tag{8.12}$$

where m_1 and m_2 are defined in Equation (8.11).

Seely and Lee [60] provided simulations that indicate Equation (8.12) has a confidence coefficient of less than the stated level in some situations. These situations occur when there is great variability among the n_q values. For example, 95% intervals had actual confidence coefficients as low as 84% in one of the designs simulated by Seely and Lee. In this design, the values of the n_q used to compute m_1 were $n_1 = 3$ and $n_2 = 72$. In a second design where $n_1 = n_2 = 4$, all the simulated confidence coefficients were at least 95%. If conditions are favorable for using the Satterthwaite approximation (i.e., if the n_q are of comparable size), then Equation (8.12) will generally provide shorter intervals than either Equation (8.8) or the GCI.

In summary, if conditions are favorable for the Satterthwaite approximation of $\widehat{\nu_1}$ and $\widehat{\nu_2}$, we recommend the interval shown in Equation (8.12). If the degrees of freedom used to compute either m_1 or m_2 vary a great deal, we recommend either Equation (8.8) or the GCI because these intervals better maintain the stated level of confidence. The lengths of the MLS intervals and GCIs are generally comparable.

8.2.3 MLS Intervals for γ_3

We begin by writing γ_3 as

$$\gamma_3 = \frac{\nu_1}{\nu_2},$$

where ν_1 and ν_2 are defined in Equation (8.9). Cochran [16] proposed the $100(1-\alpha)\%$ approximate interval for γ_3,

$$L = \frac{\widehat{\nu_1}}{\widehat{\nu_2} F_{1-\alpha/2:m_1,m_2}}$$

and

$$U = \frac{\widehat{\nu}_1}{\widehat{\nu}_2 F_{\alpha/2:m_1,m_2}}, \tag{8.13}$$

where $\widehat{\nu}_1$ and $\widehat{\nu}_2$ are defined in Equation (8.10) and m_1 and m_2 are defined in Equation (8.11). This approximation works well when the degrees of freedom are of similar magnitude. However, when the degrees of freedom vary a great deal, Equation (8.13) can yield a confidence coefficient of much less than the stated level.

An MLS interval that generally maintains the confidence level across all conditions was proposed by Ting, Burdick, and Graybill [63]. This $100(1-\alpha)\%$ interval for γ_3 is

$$L = \frac{\widehat{\nu}_1}{\widehat{\nu}_2}\left[\frac{2 + k_2/(\widehat{\nu}_1\widehat{\nu}_2) - \sqrt{V_L}}{2(1 - k_3/\widehat{\nu}_2^2)}\right]$$

and

$$U = \frac{\widehat{\nu}_1}{\widehat{\nu}_2}\left[\frac{2 + k_5/(\widehat{\nu}_1\widehat{\nu}_2) + \sqrt{V_U}}{2(1 - k_6/\widehat{\nu}_2^2)}\right], \tag{8.14}$$

where

$$\begin{aligned}
V_L &= [2 + k_2/(\widehat{\nu}_1\widehat{\nu}_2)]^2 - 4(1 - k_3/\widehat{\nu}_2^2)(1 - k_1/\widehat{\nu}_1^2),\\
V_U &= [2 + k_5/(\widehat{\nu}_1\widehat{\nu}_2)]^2 - 4(1 - k_6/\widehat{\nu}_2^2)(1 - k_4/\widehat{\nu}_1^2),\\
k_1 &= \sum_{r=1}^{P} G_r^2 d_r^2 S_r^4 + \sum_{r=1}^{P-1}\sum_{u>r}^{P} G_{ru}^* d_r d_u S_r^2 S_u^2,\\
k_2 &= \sum_{r=1}^{P}\sum_{t=P+1}^{Q} G_{rt} d_r e_t S_r^2 S_t^2,\\
k_3 &= \sum_{t=P+1}^{Q} H_t^2 e_t^2 S_t^4,\\
k_4 &= \sum_{r=1}^{P} H_r^2 d_r^2 S_r^4,\\
k_5 &= \sum_{r=1}^{P}\sum_{t=P+1}^{Q} H_{rt} d_r e_t S_r^2 S_t^2,\\
k_6 &= \sum_{t=P+1}^{Q} G_t^2 e_t^2 S_t^4 + \sum_{t=P+1}^{Q-1}\sum_{v>t}^{Q} H_{tv}^* e_t e_v S_t^2 S_v^2,
\end{aligned}$$

$\widehat{\nu}_1$ and $\widehat{\nu}_2$ are defined in Equation (8.10), and G_r, H_t, G_{rt}, G_{ru}^*, H_r, G_t, H_{rt}, and H_{tv}^* are defined in Equation (8.8).

A simpler closed-form interval was proposed by Lu, Graybill, and Burdick [45] for the special case where $P = 2$ and $Q = 3$. This is the method used to derive Equation (3.12) in Chapter 3. The performance of Equation (8.14) is comparable to the GCI for γ_3. Both of these intervals maintain the stated confidence level better than Equation (8.13).

The exact interval for the special case of γ_3 where $P = 1$ and $Q = 2$ is

$$L = \frac{d_1 S_1^2}{e_2 S_2^2 F_{1-\alpha/2:n_1,n_2}}$$

and

$$U = \frac{d_1 S_1^2}{e_2 S_2^2 F_{\alpha/2:n_1,n_2}}.$$

8.2.4 MLS Intervals for γ_4

The parameter γ_4 can be written using previous notation as

$$\gamma_4 = \frac{\nu_1 - \nu_2}{\gamma_1},$$

where ν_1 and ν_2 are defined in Equation (8.9).

Two special cases of γ_4 were encountered in previous chapters. The interval for γ_R in Equation (3.5) is of the form

$$\gamma_4 = \frac{d_1\theta_1 - e_3\theta_3}{c_2\theta_2 + c_3\theta_3 + c_4\theta_4}. \tag{8.15}$$

Confidence intervals for γ_4 in Equation (8.15) were proposed by Leiva and Graybill [42]. Arteaga, Jeyaratnam, and Graybill [2] provided an interval for the special form

$$\gamma_4 = \frac{d_1\theta_1 - e_3\theta_3}{c_2\theta_2 + c_3\theta_3}. \tag{8.16}$$

We used these results to construct the interval for γ_R in Equation (5.5).

Gui et al. [26] described a general MLS strategy for constructing a confidence interval on γ_4. However, this approach can be quite complex and requires solution of a quadratic equation. As a more practical alternative, we recommend a strategy that combines the Satterthwaite approximation with an MLS interval recommended by Wang and Graybill [70] for the ratio $(d_1\theta_1 - e_2\theta_2)/\theta_3 \geq 0$. This approximate $100(1-\alpha)\%$ interval for γ_4 is

$$L = \frac{\widehat{\nu}_2}{\widehat{\gamma}_1 F_{1-\alpha/2:m_1,m}}\left[\frac{\widehat{\nu}_1}{\widehat{\nu}_2} - F_{1-\alpha/2:m_1,\infty} + \frac{\widehat{\nu}_2 F_{1-\alpha/2:m_1,m_2}(F_{1-\alpha/2:m_1,\infty} - F_{1-\alpha/2:m_1,m_2})}{\widehat{\nu}_1}\right]$$

and

$$U = \frac{\widehat{\nu}_2}{\widehat{\gamma}_1 F_{\alpha/2:m_1,m}}\left[\frac{\widehat{\nu}_1}{\widehat{\nu}_2} - F_{\alpha/2:m_1,\infty} + \frac{\widehat{\nu}_2 F_{\alpha/2:m_1,m_2}(F_{\alpha/2:m_1,\infty} - F_{\alpha/2:m_1,m_2})}{\widehat{\nu}_1}\right], \tag{8.17}$$

where $\widehat{\nu}_1$ and $\widehat{\nu}_2$ are defined in Equation (8.10), m_1 and m_2 are defined in Equation (8.11), m is defined in Equation (8.4), and $\widehat{\gamma}_1$ is defined in Equation (8.3).

This approach was used by Adamec and Burdick [1] to construct a confidence interval for SNR in a balanced three-factor crossed random model. These results are presented in Section 8.4. If it is not known whether $(d_1\theta_1 - e_2\theta_2)/\theta_3 \geq 0$, a modified form of Equation (8.17) can be applied based on the work of Lu, Graybill, and Burdick [46, 47] as described in [10, pp. 49–51].

8.2.5 Modification for Mixed Models

In the balanced mixed model, the mean squares associated with the fixed effects have scaled noncentral chi-squared distributions (see, e.g., Hocking [31, p. 460]). Thus, one can approximate the distribution of these mean squares with chi-squared random variables as discussed in Chapter 6.

8.2.6 Modification for Unbalanced Designs

Unweighted sums of squares and harmonic means of sample sizes can be used for unbalanced designs as described in Chapter 7. We provide more references on this approach in Section 8.3.9.

8.2.7 Summary

As noted earlier, the strategies described in this section have been used to derive the confidence interval formulas in this book. To further demonstrate these strategies, we use them to analyze three additional ANOVA designs.

8.3 A Two-Fold Nested Design

Jensen [35] described a typical measurement study in the semiconductor industry. Measurements are made of quality characteristics at multiple locations on a wafer. Wafers are processed in groups called lots, and it is desired to account for variation among lots, among wafers, and within wafers. Table 8.5 gives a partial listing of a data set reported by Jensen [35, p. 647]. In this experiment there are 20 lots, 2 wafers per lot, and 9 measurements per wafer. The response variable is coded and unlabeled for proprietary reasons. Since both lots and wafers are selected at random, the responses can be modeled as a balanced two-fold nested random design. The appropriate model is

$$Y_{ijk} = \mu_Y + L_i + W_{j(i)} + E_{ijk}, \quad (8.18)$$
$$i = 1, \ldots, l, \quad j = 1, \ldots, w, \quad k = 1, \ldots, r,$$

where μ_Y is a constant and L_i, $W_{j(i)}$, and E_{ijk} are jointly independent normal random variables with means of zero and variances σ_L^2, $\sigma_{W:L}^2$, and σ_E^2, respectively. Here wafers are

Table 8.5. *Partial listing of data for wafer study.*

		Site measurements			
Lot	Wafer	1	2	...	9
1	1	181.247	181.280	...	183.117
1	2	175.267	179.844	...	181.514
2	1	167.718	169.956	...	173.530
⋮	⋮	⋮	⋮	⋮	⋮
20	2	162.084	164.154	...	161.716

Table 8.6. *ANOVA for two-fold nested design.*

Source of variation	Degrees of freedom	Mean square	Expected mean square
Lots	$n_1 = l - 1$	S_1^2	$\theta_1 = \sigma_E^2 + r\sigma_{W:L}^2 + wr\sigma_L^2$
Wafers within lots	$n_2 = l(w - 1)$	S_2^2	$\theta_2 = \sigma_E^2 + r\sigma_{W:L}^2$
Replicates	$n_3 = lw(r - 1)$	S_3^2	$\theta_3 = \sigma_E^2$

Table 8.7. *Mean squares and means for two-fold nested design.*

Statistic	Definition
S_1^2	$\frac{wr\sum_i(\overline{Y}_{i**} - \overline{Y}_{***})^2}{n_1}$
S_2^2	$\frac{r\sum_i\sum_j(\overline{Y}_{ij*} - \overline{Y}_{i**})^2}{n_2}$
S_3^2	$\frac{\sum_i\sum_j\sum_k(Y_{ijk} - \overline{Y}_{ij*})^2}{n_3}$
$\overline{Y}_{ij*}$	$\frac{\sum_k Y_{ijk}}{r}$
$\overline{Y}_{i**}$	$\frac{\sum_j\sum_k Y_{ijk}}{wr}$
$\overline{Y}_{***}$	$\frac{\sum_i\sum_j\sum_k Y_{ijk}}{lwr}$

nested within lots, and replicates are nested within wafers. The ANOVA for model (8.18) is shown in Table 8.6, and means and mean squares are defined in Table 8.7. We have subscripted the mean squares and expected mean squares with numbers to correspond with the general notation in Section 8.2. Table 8.8 reports distributional results based on the assumptions of model (8.18).

Variation of the manufacturing process is due to both lots and wafers, whereas variation of the measurement system is due only to the replication error. Table 8.9 defines the gauge R&R parameters in general notation where

$$\gamma_2 = d_1\theta_1 + d_2\theta_2 - e_3\theta_3,$$

$$\gamma_3 = \frac{d_1\theta_1 + d_2\theta_2}{e_3\theta_3},$$

$$d_1 = \frac{1}{wr},$$

$$d_2 = \frac{w - 1}{wr},$$

Table 8.8. *Distributional results for model* (8.18).

Result	
1	$\overline{Y}_{***}$, S_1^2, S_2^2, and S_3^2 are jointly independent.
2	$\dfrac{n_1 S_1^2}{\theta_1}$ is a chi-squared random variable with n_1 degrees of freedom.
3	$\dfrac{n_2 S_2^2}{\theta_2}$ is a chi-squared random variable with n_2 degrees of freedom.
4	$\dfrac{n_3 S_3^2}{\theta_3}$ is a chi-squared random variable with n_3 degrees of freedom.
5	$\overline{Y}_{***}$ is a normal random variable with mean μ_Y and variance $\dfrac{\theta_1}{lwr}$.

Table 8.9. *Gauge R&R parameters in two-fold nested design.*

Gauge R&R notation	Model (8.18) representation	Table 8.2 representation
γ_P	$\sigma_L^2 + \sigma_{W:L}^2$	γ_2
γ_M	σ_E^2	θ_3
γ_R	$\dfrac{\sigma_L^2 + \sigma_{W:L}^2}{\sigma_E^2}$	$e_3(\gamma_3 - 1)$

and

$$e_3 = \frac{1}{r}. \tag{8.19}$$

We now present confidence intervals for the gauge R&R parameters using the general strategies described in Section 8.2. Constants used in the confidence intervals are defined in Table 8.10.

8.3.1 Interval for μ_Y

The confidence interval for μ_Y is based on Results 1, 2, and 5 of Table 8.8. Using these results,

$$\frac{\sqrt{lwr}(\overline{Y}_{***} - \mu_Y)}{\sqrt{S_1^2}}$$

Table 8.10. *Constants used in confidence intervals for model* (8.18). *Values are for* $\alpha = 0.05$, $l = 20$, $w = 2$, *and* $r = 9$.

Constant	Definition	Value
G_1	$1 - F_{\alpha/2:\infty,n_1}$	0.4217
G_2	$1 - F_{\alpha/2:\infty,n_2}$	0.4147
G_3	$1 - F_{\alpha/2:\infty,n_3}$	0.1385
G_C	$1 - F_{\alpha/2:\infty,n_1+n_2}$	0.3290
H_1	$F_{1-\alpha/2:\infty,n_1} - 1$	1.133
H_2	$F_{1-\alpha/2:\infty,n_2} - 1$	1.085
H_3	$F_{1-\alpha/2:\infty,n_3} - 1$	0.1750
F_1	$F_{1-\alpha/2:n_1,n_3}$	1.772
F_2	$F_{\alpha/2:n_1,n_3}$	0.4630
F_3	$F_{1-\alpha/2:n_2,n_3}$	1.752
F_4	$F_{\alpha/2:n_2,n_3}$	0.4734
G_{13}	$\dfrac{(F_1-1)^2 - G_1^2F_1^2 - H_3^2}{F_1}$	0.004047
H_{13}	$\dfrac{(1-F_2)^2 - H_1^2F_2^2 - G_3^2}{F_2}$	–0.01311
G_{23}	$\dfrac{(F_3-1)^2 - G_2^2F_3^2 - H_3^2}{F_3}$	0.003923
H_{23}	$\dfrac{(1-F_4)^2 - H_2^2F_4^2 - G_3^2}{F_4}$	–0.01235
G_{12}^*	$G_C^2\dfrac{[n_1+n_2]^2}{n_1n_2} - \dfrac{G_1^2n_1}{n_2} - \dfrac{G_2^2n_2}{n_1}$	0.08327

has a t-distribution with n_1 degrees of freedom. From this fact, the bounds of an exact $100(1-\alpha)\%$ confidence interval for μ_Y are

$$L = \overline{Y}_{***} - \sqrt{\frac{S_1^2 F_{1-\alpha:1,n_1}}{lwr}}$$

and

$$U = \overline{Y}_{***} + \sqrt{\frac{S_1^2 F_{1-\alpha:1,n_1}}{lwr}}. \tag{8.20}$$

The ANOVA for the complete data set represented in Table 8.5 is shown in Table 8.11. For these data, $l = 20$, $w = 2$, $r = 9$, $d_1 = 0.05556$, $d_2 = 0.05556$, and $e_3 = 0.1111$. The realized value of the overall mean is $\overline{y}_{***} = 174.3$. Using the ANOVA in Table 8.11 with $F_{0.95:1,19} = 4.3807$, the 95% confidence interval for μ_Y is from $L = 171.3$ to $U = 177.3$.

Table 8.11. *ANOVA for wafer study example.*

Source of variation	Degrees of freedom	Mean square
Lots	19	739.5
Wafers within lots	20	97.98
Replicates	320	19.03

8.3.2 Interval for γ_P

The parameter γ_P can be written in the form of γ_2 as

$$\gamma_P = \nu_1 - \nu_2, \tag{8.21}$$

where

$$\nu_1 = d_1\theta_1 + d_2\theta_2,$$
$$\nu_2 = e_3\theta_3,$$

and d_1, d_2, and e_3 are defined in Equation (8.19). The MLS interval resulting from Equation (8.8) is

$$L = \widehat{\gamma}_P - \sqrt{V_L}$$

and

$$U = \widehat{\gamma}_P + \sqrt{V_U}, \tag{8.22}$$

where

$$\begin{aligned}
\widehat{\gamma}_P &= \widehat{\nu}_1 - \widehat{\nu}_2,\\
\widehat{\nu}_1 &= d_1 S_1^2 + d_2 S_2^2,\\
\widehat{\nu}_2 &= e_3 S_3^2,\\
V_L &= G_1^2 d_1^2 S_1^4 + G_2^2 d_2^2 S_2^4 + H_3^2 e_3^2 S_3^4\\
&\quad + G_{13} d_1 e_3 S_1^2 S_3^2 + G_{23} d_2 e_3 S_2^2 S_3^2 + G_{12}^* d_1 d_2 S_1^2 S_2^2,\\
V_U &= H_1^2 d_1^2 S_1^4 + H_2^2 d_2^2 S_2^4 + G_3^2 e_3^2 S_3^4\\
&\quad + H_{13} d_1 e_3 S_1^2 S_3^2 + H_{23} d_2 e_3 S_2^2 S_3^2,
\end{aligned}$$

and the constants are defined in Table 8.10. For the ANOVA in Table 8.11 with $\alpha = 0.05$, $\widehat{\nu}_1 = 46.53$, $\widehat{\nu}_2 = 2.114$, $V_L = 324.4$, $V_U = 2202$, $L = 26.4$, and $U = 91.3$.

Using the Seely–Lee interval in Equation (8.12), an approximate $100(1-\alpha)\%$ interval for γ_P is

$$L = (\widehat{\nu}_1 - \widehat{\nu}_2 F_{1-\alpha/2:m_1,n_3}) F_{\alpha/2:\infty,m_1}$$

and

$$U = (\widehat{\nu}_1 - \widehat{\nu}_2 F_{\alpha/2:m_1,n_3}) F_{1-\alpha/2:\infty,m_1}, \tag{8.23}$$

where

$$m_1 = \frac{\widehat{\nu}_1^2}{d_1^2 S_1^4/n_1 + d_2^2 S_2^4/n_2}$$

and $\widehat{\nu}_1$ and $\widehat{\nu}_2$ are defined in Equation (8.22). The degrees of freedom used to determine m_1 are $n_1 = 19$ and $n_2 = 20$. Since these are close in magnitude, the Satterthwaite approximation should work well and Equation (8.23) can be recommended. The computed 95% confidence interval using Equation (8.23) with $m_1 = 23$ (truncated) is from $L = 25.9$ to $U = 89.5$. This interval is only slightly shorter than the one computed with Equation (8.22).

8.3.3 Interval for γ_M

The parameter $\gamma_M = \sigma_E^2 = \theta_E$. From Equation (8.5), the exact $100(1-\alpha)\%$ confidence interval for γ_M is

$$L = (1 - G_3)S_3^2$$

and

$$U = (1 + H_3)S_3^2, \tag{8.24}$$

where G_3 and H_3 are defined in Table 8.10. For our example, the 95% confidence interval for γ_M is from $L = 16.4$ to $U = 22.4$.

Jensen [35] did not report specification limits for the example data. However, some reasonable limits that yield a C_p value of about 1.5 are $LSL = 144$ and $USL = 204$. Using these specification limits, the 95% confidence interval for PTR is from $L = 6\sqrt{16.4}/60 = 0.405$ to $U = 6\sqrt{22.4}/60 = 0.473$.

8.3.4 Interval for γ_R

The parameter γ_R can be written as

$$\gamma_R = e_3(\gamma_3 - 1), \tag{8.25}$$

where

$$\gamma_3 = \frac{\nu_1}{\nu_2}$$

and ν_1 and ν_2 are defined in Equation (8.21). Once a $100(1-\alpha)\%$ coefficient for γ_3 is computed, it is transformed into an interval for γ_R using the relationship in Equation (8.25).

The Cochran interval defined in Equation (8.13) provides the $100(1-\alpha)\%$ interval for γ_3,

$$L = \frac{\widehat{\nu}_1}{\widehat{\nu}_2 F_{1-\alpha/2:m_1,n_3}}$$

and

$$U = \frac{\widehat{\nu}_1}{\widehat{\nu}_2 F_{\alpha/2:m_1,n_3}}, \tag{8.26}$$

where $\widehat{\nu}_1$ and $\widehat{\nu}_2$ are defined in Equation (8.22) and m_1 is defined in Equation (8.23). This interval can be recommended for our example because the degrees of freedom used in m_1 are close in magnitude ($n_1 = 19$ and $n_2 = 20$). The computed 95% confidence interval for γ_3 using Equation (8.26) uses $m_1 = 23$ (truncated), $F_{0.975:23,320} = 1.6998$, and $F_{0.025:23,320} =$

0.50101 to yield $L = 12.9$ and $U = 43.9$. The transformed 95% confidence interval for γ_R using Equation (8.25) is from $L = (0.1111)(12.9 - 1) = 1.32$ to $U = (0.1111)(43.9 - 1) = 4.77$.

The $100(1-\alpha)\%$ interval for γ_3 using the MLS interval in Equation (8.14) with $Q = 3$ and $P = 2$ is

$$L = \frac{\widehat{\upsilon}_1}{\widehat{\upsilon}_2}\left[\frac{2 + k_2/(\widehat{\upsilon}_1\widehat{\upsilon}_2) - \sqrt{V_L}}{2(1 - k_3/\widehat{\upsilon}_2^2)}\right]$$

and

$$U = \frac{\widehat{\upsilon}_1}{\widehat{\upsilon}_2}\left[\frac{2 + k_5/(\widehat{\upsilon}_1\widehat{\upsilon}_2) + \sqrt{V_U}}{2(1 - k_6/\widehat{\upsilon}_2^2)}\right], \qquad (8.27)$$

where

$$\begin{aligned}
V_L &= [2 + k_2/(\widehat{\upsilon}_1\widehat{\upsilon}_2)]^2 - 4(1 - k_3/\widehat{\upsilon}_2^2)(1 - k_1/\widehat{\upsilon}_1^2),\\
V_U &= [2 + k_5/(\widehat{\upsilon}_1\widehat{\upsilon}_2)]^2 - 4(1 - k_6/\widehat{\upsilon}_2^2)(1 - k_4/\widehat{\upsilon}_1^2),\\
k_1 &= G_1^2 d_1^2 S_1^4 + G_2^2 d_2^2 S_2^4 + G_{12}^* d_1 d_2 S_1^2 S_2^2,\\
k_2 &= G_{13} d_1 e_3 S_1^2 S_3^2 + G_{23} d_2 e_3 S_2^2 S_3^2,\\
k_3 &= H_3^2 e_3^2 S_3^4,\\
k_4 &= H_1^2 d_1^2 S_1^4 + H_2^2 d_2^2 S_2^4,\\
k_5 &= H_{13} d_1 e_3 S_1^2 S_3^2 + H_{23} d_2 e_3 S_2^2 S_3^2,\\
k_6 &= G_3^2 e_3^2 S_3^4,
\end{aligned}$$

$\widehat{\upsilon}_1$ and $\widehat{\upsilon}_2$ are defined in Equation (8.22), and the constants are defined in Table 8.10. The computed 95% interval on γ_3 based on Equation (8.27) uses $k_1 = 323.8$, $k_2 = 0.3967$, $k_3 = 0.1369$, $k_4 = 2203$, $k_5 = -1.281$, $k_6 = 0.08578$, $V_L = 0.7187$, and $V_U = 4.017$ to provide $L = 13.1$ and $U = 44.8$. The transformed 95% confidence interval for γ_R using Equation (8.25) is from $L = (0.1111)(13.1 - 1) = 1.34$ to $U = (0.1111)(44.8 - 1) = 4.87$.

Based on the interval for γ_R using Equation (8.26), the 95% confidence interval for SNR is from $L = \sqrt{2(1.32)} = 1.6$ to $U = \sqrt{2(4.77)} = 3.1$.

8.3.5 Generalized Confidence Intervals

The GPQs for model (8.18) are shown in Table 8.12, where s_1^2, s_2^2, and s_3^2 are the realized values of S_1^2, S_2^2, and S_3^2 and W_1, W_2, and W_3 are independent chi-squared random variables with n_1, n_2, and n_3 degrees of freedom. They are based on the general GPQ functions in Table 8.4. No GPQs are needed for μ_Y and γ_M since exact intervals are provided in Equations (8.20) and (8.24), respectively. Table 8.13 shows the computed 95% confidence intervals for our example based on 100,000 computed GPQs. For comparison, we have included the MLS intervals. Both sets of intervals are comparable for this example.

8.3.6 Intervals for Misclassification Rates

Using the specification limits $LSL = 144$ and $USL = 204$ with $\epsilon = 0.001$, confidence intervals for the misclassification rates are computed using the GPQs shown in Table 8.14.

Table 8.12. *GPQs for model* (8.18).

Parameter	GPQ
γ_P	$\max\left[0, \dfrac{d_1 n_1 s_1^2}{W_1} + \dfrac{d_2 n_2 s_2^2}{W_2} - \dfrac{e_3 n_3 s_3^2}{W_3}\right]$
γ_R	$\dfrac{\text{GPQ}(\gamma_P)}{n_3 s_3^2 / W_3}$

Table 8.13. *Comparison of intervals for wafer study example. (See Section* 1.8 *for a description of computer programs to perform these computations.)*

Parameter	MLS	Generalized
γ_P	$L = 26.4$ $U = 91.3$	$L = 27.1$ $U = 91.9$
γ_R	$L = 1.34$ $U = 4.87$	$L = 1.39$ $U = 4.89$

Table 8.14. *GPQs for misclassification rates in model* (8.18).

Parameter	GPQ
$\mu_Y = \mu_P$	$\overline{y}_{***} - Z\sqrt{\dfrac{n_1 s_1^2}{lwrW_1}}$
$\gamma_P + \gamma_M$	$\dfrac{n_1 s_1^2}{wrW_1} + \dfrac{(w-1)n_2 s_2^2}{wrW_2} + \dfrac{(r-1)n_3 s_3^2}{rW_3}$
γ_P	$\max\left[\epsilon, \dfrac{d_1 n_1 s_1^2}{W_1} + \dfrac{d_2 n_2 s_2^2}{W_2} - \dfrac{e_3 n_3 s_3^2}{W_3}\right]$

Table 8.15. 95% *confidence intervals for misclassification rates. (Bounds for δ, δ_c, β, and β_c have been multiplied by* 10^6.*)*

Parameter	Lower bound	Upper bound
δ	12	3,768
δ_c	12	3,777
δ_{index}	1.65	941
β	0.006	671
β_c	299,145	422,913
β_{index}	0.300	0.423

The resulting intervals for our example are shown in Table 8.15. It is clear from the interval for δ_{index} that the system is not capable of discriminating parts. Although the intervals for δ and β contain relatively small values, the system is not capable because the process has such

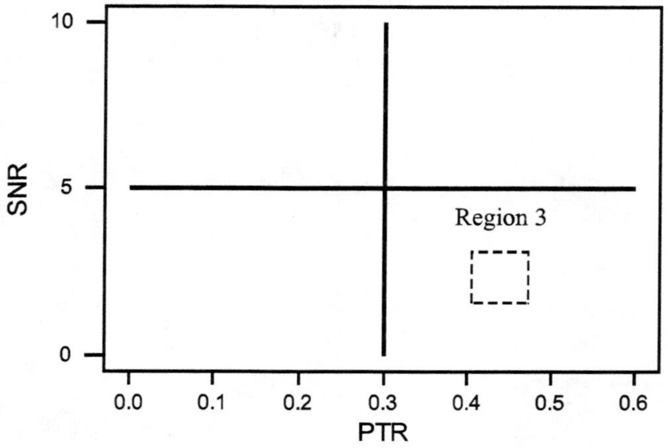

Figure 8.1. *R&R graph for wafer study.*

a high capability measure. (The 95% MLS confidence interval for C_p is from $L = 1.05$ to $U = 1.95$.)

8.3.7 Conclusions

The R&R graph in Figure 8.1 indicates that the measurement system satisfies neither of the capability criteria. That all values in the confidence interval for δ_{index} are greater than one also suggests the system is not capable.

The residual plots in Figure 8.2 indicate a slight departure from normality based on the normal probability plot and the histogram. There could also be a problem with nonconstant variance as indicated by the residuals versus the fitted values plot. In particular, the residuals appear to increase as the fitted values increase. If the departures are considered extreme, then transformation of the data may be in order to correct the problems. However, ANOVA techniques are known to be robust to slight departures from the underlying assumptions. In this example, the departures do not appear to be extreme enough to change our conclusion that the measurement system is not capable.

Figure 8.3 displays the $\overline{X}$ and R control charts. These plots are consistent with the conclusions drawn from the PTR and SNR criteria. In particular, all of the sample means except one in the $\overline{X}$ control chart are within the specification limits. This implies that the measurement system cannot discriminate between parts. The R control chart shows increased variation in the data over time (assuming the parts are ordered sequentially).

8.3.8 Other MLS Intervals

Burdick and Graybill [10, pp. 78–91] provide MLS intervals for other parameters in model (8.18) not typically computed in a gauge R&R study. These parameters include σ_L^2, $\sigma_{W:L}^2$, σ_L^2/σ_E^2, $\sigma_{W:L}^2/\sigma_E^2$, σ_L^2/γ_Y, $\sigma_{W:L}^2/\gamma_Y$, and σ_E^2/γ_Y, where $\gamma_Y = \gamma_P + \gamma_M$. These intervals are based on the general MLS strategies discussed in Section 8.2 as well as some special-case strategies for the two-fold nested design (see, e.g., Graybill and Wang [24]).

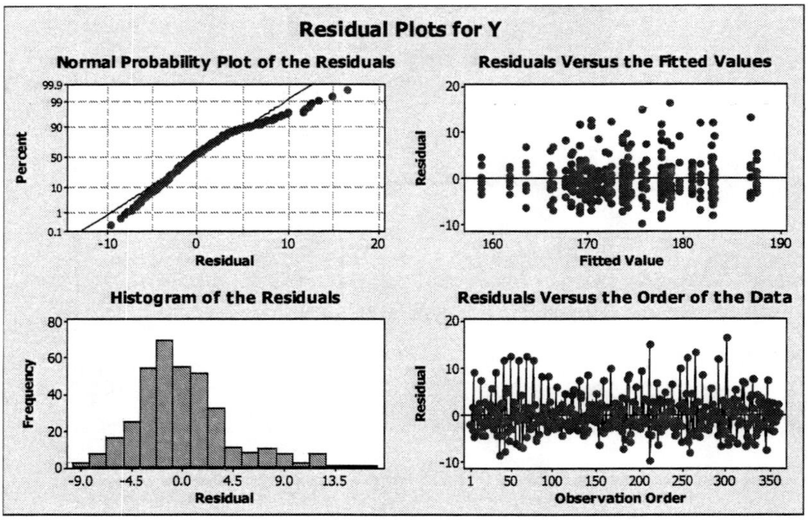

Figure 8.2. *Residual plots for wafer study.*

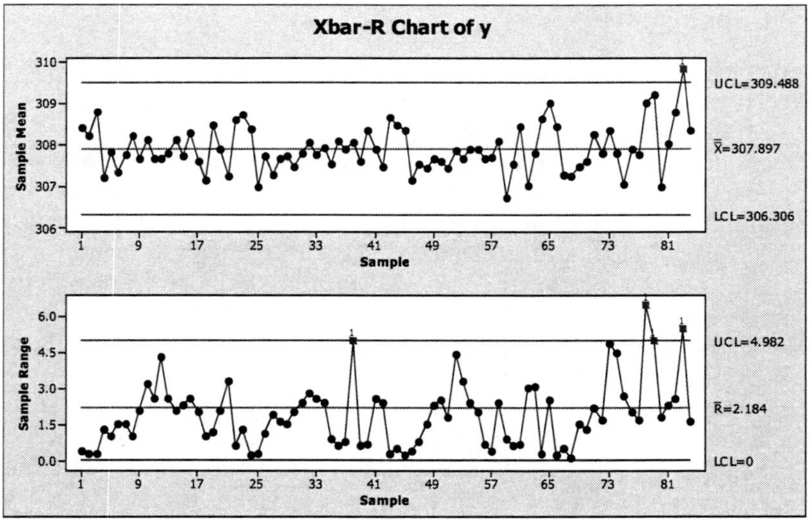

Figure 8.3. $\overline{X}$ *and R control charts for wafer study.*

8.3.9 Unbalanced Modifications

A nested design can have unequal replications at any level of nesting. Burdick and Graybill [10, p. 99] provided formulas for unweighted sums of squares and harmonic means in an unbalanced two-fold nested design. Hernandez, Burdick, and Birch [30] and Burdick, Birch, and Graybill [7] demonstrated that these approximations work well and provide confidence intervals that are generally close to the stated level.

8.4 Three-Factor Crossed Random Designs

In Chapters 3 and 5 we presented results for a two-factor crossed random design. We now present some results for a three-factor crossed random design.

Table 8.16 presents a partial listing of data from a gauge R&R study used in the fabrication of magnetic tape for computerized data storage. In this study, o automated test stations (operators) are used to evaluate the quality of p tape heads (parts). To measure characteristics of the heads, t tapes are used in each test station. In this design, all p heads are measured with each of the ot test station × tape combinations. There are $r = 2$ replicates for each treatment condition. The response variable is "Forward resolution" of a tape head. In the process of writing information to magnetic tape, voltage is measured at two different frequencies. The resolution is a ratio of these two voltages. There is no unit of measurement on the ratio, and it is expressed as a percentage. Ideally, this ratio should be 100%. The specification limits are $LSL = 90\%$ and $USL = 110\%$. Forward resolution is measured when the drive writes in the forward direction.

The model used to represent the three-factor random design is

$$\begin{aligned} Y_{ijkl} = \mu_Y &+ P_i + O_j + T_k + (PO)_{ij} \\ &+ (PT)_{ik} + (OT)_{jk} + (POT)_{ijk} + E_{ijkl}, \\ i = 1, \ldots, p, \quad & j = 1, \ldots, o, \quad k = 1, \ldots, t, \quad l = 1, \ldots, r, \end{aligned} \tag{8.28}$$

where μ_Y is a constant and P_i, O_j, T_k, $(PO)_{ij}$, $(PT)_{ik}$, $(OT)_{jk}$, $(POT)_{ijk}$, and E_{ijkl} are

Table 8.16. *Example data for a three-factor crossed design.*

Part	Operator	Tape	Y
1	1	1	102.8917
1	1	1	103.3077
1	1	2	104.1766
1	1	2	103.9411
1	1	3	103.6517
1	1	3	103.8112
1	2	1	102.4783
1	2	1	103.3801
⋮	⋮	⋮	⋮
9	3	1	97.08995
9	3	1	96.94569
9	3	2	97.28612
9	3	2	97.72192
9	3	3	96.66787
9	3	3	96.74137

Table 8.17. *ANOVA for three-factor crossed design.*

Source of variation	Degrees of freedom	Mean square	Expected mean square
Parts (P)	$n_1 = p - 1$	S_1^2	$\theta_1 = \sigma_E^2 + r\sigma_{POT}^2 + or\sigma_{PT}^2 + tr\sigma_{PO}^2$ $+otr\sigma_P^2$
Operators (O)	$n_2 = o - 1$	S_2^2	$\theta_2 = \sigma_E^2 + r\sigma_{POT}^2 + tr\sigma_{PO}^2 + pr\sigma_{OT}^2$ $+ptr\sigma_O^2$
Tapes (T)	$n_3 = t - 1$	S_3^2	$\theta_3 = \sigma_E^2 + r\sigma_{POT}^2 + pr\sigma_{OT}^2 + or\sigma_{PT}^2$ $+por\sigma_T^2$
PO	$n_4 = (p-1)(o-1)$	S_4^2	$\theta_4 = \sigma_E^2 + r\sigma_{POT}^2 + tr\sigma_{PO}^2$
PT	$n_5 = (p-1)(t-1)$	S_5^2	$\theta_5 = \sigma_E^2 + r\sigma_{POT}^2 + or\sigma_{PT}^2$
OT	$n_6 = (o-1)(t-1)$	S_6^2	$\theta_6 = \sigma_E^2 + r\sigma_{POT}^2 + pr\sigma_{OT}^2$
POT	$n_7 = (p-1)(o-1)(t-1)$	S_7^2	$\theta_7 = \sigma_E^2 + r\sigma_{POT}^2$
Replicates	$n_8 = pot(r-1)$	S_8^2	$\theta_8 = \sigma_E^2$

jointly independent normal random variables with means of zero and variances $\sigma_P^2, \sigma_O^2, \sigma_T^2, \sigma_{PO}^2, \sigma_{PT}^2, \sigma_{OT}^2, \sigma_{POT}^2$, and σ_E^2, respectively.

The ANOVA for model (8.28) is shown in Table 8.17. The overall mean

$$\overline{Y}_{****} = \frac{\sum_{i=1}^{p}\sum_{j=1}^{o}\sum_{k=1}^{t}\sum_{l=1}^{r} Y_{ijkl}}{potr}$$

is a normal random variable with mean μ_Y and variance

$$\frac{\theta_1 + \theta_2 + \theta_3 + \theta_7 - (\theta_4 + \theta_5 + \theta_6)}{potr}.$$

Because the design is balanced, $n_q S_q^2/\theta_q$ has a chi-squared distribution with n_q degrees of freedom ($q = 1, \ldots, 8$), and all mean squares are jointly independent. The computed ANOVA for the complete data set represented in Table 8.16 is shown in Table 8.18.

Variation of the manufacturing process is represented by the part variation. Variation from operators, tapes, and all interactions is attributed to the measurement system. Thus, $\gamma_P = \sigma_P^2$ and $\gamma_M = \sigma_O^2 + \sigma_T^2 + \sigma_{PO}^2 + \sigma_{PT}^2 + \sigma_{OT}^2 + \sigma_{POT}^2 + \sigma_E^2$. We now demonstrate the general strategy of Section 8.2 by computing confidence intervals for γ_P, γ_M, and $\gamma_R = \gamma_P/\gamma_M$.

Table 8.18. *ANOVA for three-factor example.*

Source of variation	Degrees of freedom	Mean square
Parts (P)	8	102.8
Operators (O)	2	13.08
Tapes (T)	2	29.23
PO	16	1.074
PT	16	1.884
OT	4	2.932
POT	32	0.4183
Replicates	81	0.1394

8.4.1 Interval for γ_P

The parameter $\gamma_P = \sigma_P^2$ is written in terms of the expected mean squares as

$$\gamma_P = \nu_1 - \nu_2,$$

where

$$\nu_1 = \frac{\theta_1 + \theta_7}{otr}$$

and

$$\nu_2 = \frac{\theta_4 + \theta_5}{otr}.$$

Note that γ_P is in the form of γ_2 as defined in Table 8.2. Strategies for constructing intervals for γ_2 are discussed in Section 8.2.2. For the example data set, we recommend the Seely–Lee interval in Equation (8.12). This interval is recommended because the degrees of freedom used for m_2 are equal ($n_4 = n_5 = 16$), and the minimum value of the degrees of freedom used for m_1 ($n_1 = 8$) is large enough for the Satterthwaite approximation to work well. The resulting $100(1-\alpha)\%$ interval for γ_P is

$$L = (\widehat{\nu}_1 - \widehat{\nu}_2 F_{1-\alpha/2:m_1,m_2}) F_{\alpha/2:\infty,m_1}$$

and

$$U = (\widehat{\nu}_1 - \widehat{\nu}_2 F_{\alpha/2:m_1,m_2}) F_{1-\alpha/2:\infty,m_1}, \tag{8.29}$$

where

$$\widehat{\nu}_1 = \frac{S_1^2 + S_7^2}{otr},$$

$$\widehat{\nu}_2 = \frac{S_4^2 + S_5^2}{otr},$$

$$m_1 = \frac{(S_1^2 + S_7^2)^2}{S_1^4/n_1 + S_7^4/n_7},$$

and

$$m_2 = \frac{(S_4^2 + S_5^2)^2}{S_4^4/n_4 + S_5^4/n_5}.$$

The computed 95% interval for γ_P with $\widehat{\nu}_1 = 5.737$, $\widehat{\nu}_2 = 0.1643$, $m_1 = 8$ (truncated), $m_2 = 29$ (truncated), $F_{0.975:8,29} = 2.6686$, $F_{0.975:8,\infty} = 2.1918$, $F_{0.025:8,29} = 0.2563$, and $F_{.025:8,\infty} = 0.2725$ is from $L = 2.42$ to $U = 20.9$.

The GPQ for γ_P is

$$\max\left[0, \frac{n_1 s_1^2}{otr W_1} + \frac{n_7 s_7^2}{otr W_7} - \frac{n_4 s_4^2}{otr W_4} - \frac{n_5 s_5^2}{otr W_5}\right], \tag{8.30}$$

where s_1^2, s_4^2, s_5^2, and s_7^2 are the realized values of S_1^2, S_4^2, S_5^2, and S_7^2 and W_1, W_4, W_5, and W_7 are random independent chi-squared random variables with n_1, n_4, n_5, and n_7 degrees of freedom, respectively. The computed 95% generalized confidence interval for γ_P based on 100,000 simulated GPQ values is from $L = 2.46$ to $U = 20.9$.

8.4.2 Interval for γ_M

The parameter $\gamma_M = \sigma_O^2 + \sigma_T^2 + \sigma_{PO}^2 + \sigma_{PT}^2 + \sigma_{OT}^2 + \sigma_{POT}^2 + \sigma_E^2$ can be written in the form of γ_1 in Table 8.2. In particular,

$$\gamma_M = \frac{\sum_{q=2}^{8} c_q \theta_q}{potr}, \tag{8.31}$$

where $c_2 = o$, $c_3 = t$, $c_4 = (p-1)o$, $c_5 = (p-1)t$, $c_6 = ot - o - t$, $c_7 = o + t - po - pt - ot + pot$, and $c_8 = pot(r-1)$.

Methods for constructing confidence intervals on γ_1 are discussed in Section 8.2.1. Since $\min(n_1, n_2, \ldots, n_8) = n_2 = n_3 = 2$, and the other degrees of freedom vary greatly in magnitude, the Satterthwaite interval shown in Equation (8.4) cannot be recommended. Instead, we use the Graybill–Wang formulation in Equation (8.3). This provides the 95% confidence interval for γ_M from $L = 0.950$ to $U = 24.2$. The resulting 95% interval for PTR using $LSL = 90$, $USL = 110$, and $k = 6$ is $L = 6\sqrt{0.950}/20 = 0.292$ and $U = 6\sqrt{24.2}/20 = 1.48$.

The GPQ for γ_M is

$$\sum_{q=2}^{8} \frac{c_q n_q s_q^2}{potr W_q}, \tag{8.32}$$

where the c_q are defined in Equation (8.31), $s_2^2, \ldots, s_8^2$ are the realized values of $S_2^2, \ldots, S_8^2$, and $W_2, \ldots, W_8$ are random independent chi-squared random variables with $n_2, \ldots, n_8$ degrees of freedom, respectively. The computed 95% GCI for γ_M based on 100,000 simulated GPQ values is from $L = 1.00$ to $U = 33.3$.

8.4.3 Interval for γ_R

The parameter $\gamma_R = \gamma_P/\gamma_M = (\nu_1 - \nu_2)/\gamma_M$ is of the form γ_4 in Table 8.2. Section 8.2.4 describes how to construct intervals for γ_4. Using Equation (8.17), an approximate

$100(1-\alpha)\%$ interval for γ_R is

$$L = \frac{\widehat{\nu}_2}{\widehat{\gamma}_M F_{1-\alpha/2:m_1,m}} \left[\frac{\widehat{\nu}_1}{\widehat{\nu}_2} - F_{1-\alpha/2:m_1,\infty} + \frac{\widehat{\nu}_2 F_{1-\alpha/2:m_1,m_2}(F_{1-\alpha/2:m_1,\infty} - F_{1-\alpha/2:m_1,m_2})}{\widehat{\nu}_1} \right]$$

and

$$U = \frac{\widehat{\nu}_2}{\widehat{\gamma}_M F_{\alpha/2:m_1,m}} \left[\frac{\widehat{\nu}_1}{\widehat{\nu}_2} - F_{\alpha/2:m_1,\infty} + \frac{\widehat{\nu}_2 F_{\alpha/2:m_1,m_2}(F_{\alpha/2:m_1,\infty} - F_{\alpha/2:m_1,m_2})}{\widehat{\nu}_1} \right], \tag{8.33}$$

where

$$\widehat{\gamma}_M = \frac{\sum_{q=2}^{8} c_q S_q^2}{potr},$$

$$m = \frac{\left(\sum_{q=2}^{8} c_q S_q^2\right)^2}{\sum_{q=2}^{8} c_q^2 S_q^4 / n_q},$$

$\widehat{\nu}_1$, $\widehat{\nu}_2$, m_1, and m_2 are defined in Equation (8.29), and $c_2, \ldots, c_8$ are defined in Equation (8.31). The computed 95% interval for γ_R in our example with $\widehat{\nu}_1 = 5.737$, $\widehat{\nu}_2 = 0.1643$, $m_1 = 8$ (truncated), $m_2 = 29$ (truncated), $m = 10$ (truncated), and $\widehat{\gamma}_M = 1.408$ is from $L = 0.990$ to $U = 17.4$. By transforming the interval for γ_R, the 95% confidence interval for the SNR is from $L = \sqrt{2(0.990)} = 1.4$ to $U = \sqrt{2(17.4)} = 5.9$.

Adamec and Burdick [1] provided simulations that demonstrate that the interval in Equation (8.33) generally maintains the stated confidence level. The 95% GCI formed by taking the ratio of Equations (8.30) and (8.32) is from $L = 0.166$ to $U = 11.9$. This transforms into a 95% confidence interval on SNR from $L = 0.58$ to $U = 4.9$.

All of the intervals computed in this example are very wide because $n_2 = 2$ and $n_3 = 2$. These small degrees of freedom yield wide confidence intervals for any function of θ_2 or θ_3. A better design would have included more operators and tapes. It is recommended that all effects have at least four degrees of freedom in a random ANOVA model in order to provide intervals that are short enough to be useful.

8.4.4 Intervals for Misclassification Rates

The confidence intervals for the misclassification rates can be computed using the generalized inference process presented earlier in this book. The required GPQs are shown in Table 8.19, where $c_{1T} = p$, $c_{2T} = o$, $c_{3T} = t$, $c_{4T} = po - p - o$, $c_{5T} = pt - p - t$, $c_{6T} = ot - o - t$, $c_{7T} = p + o + t - po - pt - ot + pot$, and $c_{8T} = pot(r-1)$. The resulting intervals with $LSL = 90$, $USL = 110$, $\overline{y}_{****} = 100.49$, and $\epsilon = 0.001$ are shown in Table 8.20. All of the intervals are very wide because of the small number of parts, operators, and tapes used in the experiment.

8.4.5 Conclusions

The R&R graph in Figure 8.4 and confidence interval for δ_{index} in Table 8.20 do not provide evidence that the measurement system is capable.

Table 8.19. *GPQs for misclassification rates in model* (8.28).

Parameter	GPQ
$\mu_Y = \mu_P$	$\overline{y}_{****} - Z\sqrt{\max\left[\epsilon, \sum_{q=1}^{3}\frac{n_q s_q^2}{potrW_q} + \frac{n_7 s_7^2}{potrW_7} - \sum_{q=4}^{6}\frac{n_q s_q^2}{potrW_q}\right]}$
$\gamma_P + \gamma_M$	$\sum_{q=1}^{8}\frac{c_{qT} n_q S_q^2}{potrW_q}$
γ_P	$\max\left[\epsilon, \frac{n_1 s_1^2}{otrW_1} + \frac{n_7 s_7^2}{otrW_7} - \frac{n_4 s_4^2}{otrW_4} - \frac{n_5 s_5^2}{otrW_5}\right]$

Table 8.20. 95% *confidence intervals for misclassification rates. (Bounds for δ, δ_c, β, and β_c have been multiplied by* 10^6.)

Parameter	Lower bound	Upper bound
δ	3	144,994
δ_c	3	160,920
δ_{index}	0.43	22,438
β	0.0007	13,767
β_c	183,381	456,596
β_{index}	0.194	0.461

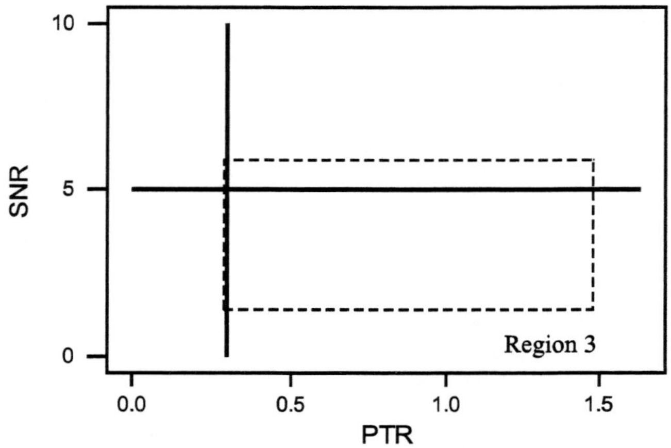

Figure 8.4. *R&R graph for three-factor example.*

Further analysis discovered a problem with the tapes used in the experiment. Figure 8.5 displays the residual plots. Examining these plots, there does not appear to be a problem with any of the underlying assumptions.

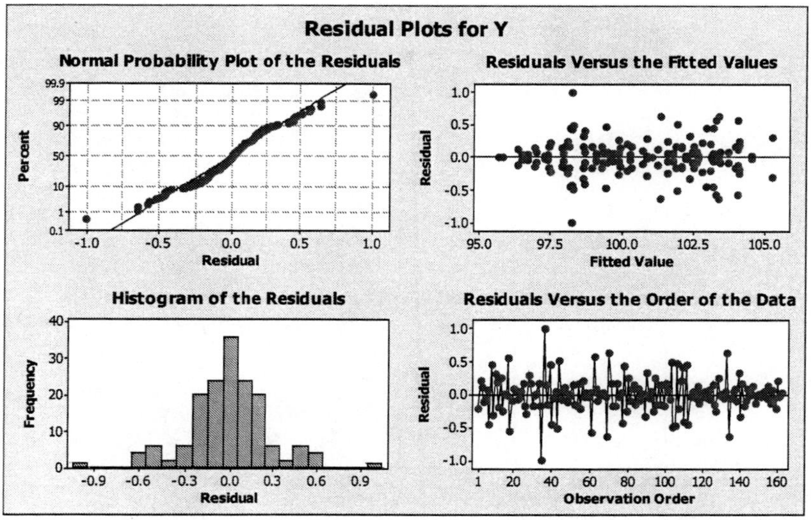

Figure 8.5. *Residual plots for three-factor example.*

Table 8.21. *Data for one day of a semiconductor study.*

Operator	Site	Measurements
1	1	308.1, 308.2, 308.6, 308.5
1	2	308.5, 308.8, 308.5, 308.8
1	3	308.5, 308.8, 308.8, 308.7
1	4	308.5, 308.6, 308.7, 308.5
2	1	308.0, 309.5, 308.8, 310.8
2	2	308.2, 308.8, 308.3, 308.6
2	3	308.3, 308.4, 308.4, 308.3
2	4	308.3, 308.2, 308.4, 308.4
3	1	308.9, 308.5, 308.4, 308.5
3	2	308.8, 308.3, 308.4, 308.5
3	3	308.8, 308.4, 308.2, 308.3
3	4	308.6, 308.3, 308.4, 308.4

8.5 Designs with Both Nested and Crossed Factors

Our final example considers a model with both crossed and nested factors. Borror, Montgomery, and Runger [4] described a semiconductor manufacturing process in which measurements are recorded at each of four sites on a wafer during the manufacturing process. Data are collected over three shifts (operators) each day for a 7-day period. Since shifts of operators differ by day, operators are nested within day. Table 8.21 displays a listing of the data for the first day. The response variable has been coded for proprietary reasons.

Table 8.22. *ANOVA for the semiconductor design.*

Source of variation	Degrees of freedom	Mean square	Expected mean square
Days (D)	$n_1 = d - 1$	S_1^2	$\theta_1 = \sigma_E^2 + r\sigma_{OS:D}^2 + sr\sigma_{O:D}^2 + osr\sigma_D^2$
Operators: Days (O:D)	$n_2 = d(o-1)$	S_2^2	$\theta_2 = \sigma_E^2 + r\sigma_{OS:D}^2 + sr\sigma_{O:D}^2$
Sites (S)	$n_3 = s - 1$	S_3^2	$\theta_3 = \sigma_E^2 + r\sigma_{OS:D}^2 + dor\sigma_S^2$
OS:D	$n_4 = (s-1)(do-1)$	S_4^2	$\theta_4 = \sigma_E^2 + r\sigma_{OS:D}^2$
Replicates	$n_5 = dos(r-1)$	S_5^2	$\theta_5 = \sigma_E^2$

The model used to represent the data in Table 8.21 is

$$Y_{ijkl} = \mu_Y + D_i + O_{j(i)} + S_k + (OS)_{jk(i)} + E_{ijkl}, \qquad (8.34)$$
$$i = 1, \ldots, d, \quad j = 1, \ldots, o, \quad k = 1, \ldots, s, \quad l = 1, \ldots, r,$$

where μ_Y is a constant and D_i, $O_{j(i)}$, S_k, $(OS)_{jk(i)}$, and E_{ijkl} are jointly independent normal random variables with means of zero and variances σ_D^2, $\sigma_{O:D}^2$, σ_S^2, $\sigma_{OS:D}^2$, and σ_E^2, respectively. The factor "day" is treated as a blocking variable, and so the Day×Site interaction is pooled with OS:D. The combined term is labeled OS:D in Table 8.22.

The ANOVA for model (8.34) is shown in Table 8.22. The overall mean

$$\overline{Y}_{****} = \frac{\sum_{i=1}^{d}\sum_{j=1}^{o}\sum_{k=1}^{s}\sum_{l=1}^{r} Y_{ijkl}}{dosr}$$

is a normal random variable with mean μ_Y and variance

$$\frac{\theta_1 + \theta_3 - \theta_4}{dosr}.$$

Because the design is balanced, $n_q S_q^2/\theta_q$ has a chi-squared distribution with n_q degrees of freedom ($q = 1, \ldots, 5$), and all mean squares are jointly independent. The ANOVA for the complete data set of 7 days is shown in Table 8.23. For the complete data set $d = 7$, $o = 3$, $s = 4$, and $r = 4$.

The variation associated with the process is $\gamma_P = \sigma_D^2 + \sigma_S^2$, and the variation associated with the measurement system is $\gamma_M = \sigma_{O:D}^2 + \sigma_{OS:D}^2 + \sigma_E^2$. We will demonstrate the general strategy in Section 8.2 by computing confidence intervals for these two parameters and $\gamma_R = \gamma_P/\gamma_M$.

8.5.1 Interval for γ_P

The parameter $\gamma_P = \sigma_D^2 + \sigma_S^2$ is written in terms of the expected mean squares as

$$\gamma_P = \nu_1 - \nu_2, \qquad (8.35)$$

Table 8.23. *Computed ANOVA for the semiconductor study.*

Source of variation	Degrees of freedom	Mean square
Days (D)	6	6.480
Operators:Days (O:D)	14	1.404
Sites (S)	3	35.22
OS:D	60	1.447
Replicates	252	0.5292

where

$$\nu_1 = \frac{d\theta_1 + s\theta_3}{dosr}$$

and

$$\nu_2 = \frac{d\theta_2 + s\theta_4}{dosr}.$$

As seen in Equation (8.35), γ_P is in the form of γ_2 in Table 8.2. We again recommend the Seely–Lee interval in Equation (8.12) for forming an interval on γ_P. The resulting $100(1-\alpha)\%$ interval for γ_P is

$$L = (\widehat{\nu}_1 - \widehat{\nu}_2 F_{1-\alpha/2:m_1,m_2}) F_{\alpha/2:\infty,m_1}$$

and

$$U = (\widehat{\nu}_1 - \widehat{\nu}_2 F_{\alpha/2:m_1,m_2}) F_{1-\alpha/2:\infty,m_1}, \tag{8.36}$$

where

$$\widehat{\nu}_1 = \frac{dS_1^2 + sS_3^2}{dosr},$$

$$\widehat{\nu}_2 = \frac{dS_2^2 + sS_4^2}{dosr},$$

$$m_1 = \frac{\widehat{\nu}_1^2 (dosr)^2}{d^2 S_1^4/n_1 + s^2 S_3^4/n_3},$$

and

$$m_2 = \frac{\widehat{\nu}_2^2 (dosr)^2}{d^2 S_2^4/n_2 + s^2 S_4^4/n_4}.$$

The computed 95% interval for γ_P with $\widehat{\nu}_1 = 0.5542$, $\widehat{\nu}_2 = 0.04647$, $m_1 = 4$ (truncated), $m_2 = 32$ (truncated), $F_{0.975:4,32} = 3.2185$, $F_{0.975:4,\infty} = 2.7858$, $F_{0.025:4,32} = 0.1184$, and $F_{.025:4,\infty} = 0.1211$ is from $L = 0.145$ to $U = 4.53$.

The GPQ for γ_P is

$$\max\left[0, \frac{n_1 s_1^2}{osr W_1} + \frac{n_3 s_3^2}{dor W_3} - \frac{n_2 s_2^2}{osr W_2} - \frac{n_4 s_4^2}{dor W_4}\right], \tag{8.37}$$

where s_1^2, s_2^2, s_3^2, and s_4^2 are the realized values of S_1^2, S_2^2, S_3^2, and S_4^2 and W_1, W_2, W_3, and W_4 are random independent chi-squared random variables with n_1, n_2, n_3, and n_4 degrees of freedom, respectively. The computed 95% generalized confidence interval for γ_P based on 100,000 simulated GPQ values is from $L = 0.205$ to $U = 6.01$.

8.5.2 Interval for γ_M

The parameter $\gamma_M = \sigma^2_{O:D} + \sigma^2_{OS:D} + \sigma^2_E$ is written in terms of the expected mean squares as

$$\gamma_M = \frac{\theta_2 + (s-1)\theta_4 + s(r-1)\theta_5}{sr}.$$

This is in the form of γ_1 and so we recommend the Graybill–Wang formulation in Equation (8.3) for constructing a confidence interval. Using Equation (8.3) with

$$\begin{aligned}\widehat{\gamma}_M &= \frac{S_2^2 + (s-1)S_4^2 + s(r-1)S_5^2}{sr} \\ &= 0.756 \end{aligned} \tag{8.38}$$

provides the 95% confidence interval from $L = 0.650$ to $U = 0.957$. Using the specification limits $LSL = 304.7$ and $USL = 311.1$, this translates into a confidence interval for PTR from $L = 0.756$ to $U = 0.917$.

The GPQ for γ_M is

$$\frac{n_2 s_2^2}{sr W_2} + \frac{(s-1)n_4 s_4^2}{sr W_4} + \frac{s(r-1)n_5 s_5^2}{sr W_5}, \tag{8.39}$$

where s_2^2, s_4^2, and s_5^2 are the realized values of S_2^2, S_4^2, S_5^2 and W_2, W_4, W_5 are random independent chi-squared random variables with n_2, n_4, and n_5 degrees of freedom, respectively. The computed 95% generalized confidence interval for γ_M based on 100,000 simulated GPQ values is from $L = 0.651$ to $U = 0.957$.

8.5.3 Interval for γ_R

The parameter $\gamma_R = \gamma_P/\gamma_M = (\nu_1 - \nu_2)/\gamma_M$ is of the form γ_4 in Table 8.2. Using Equation (8.17), an approximate $100(1-\alpha)\%$ interval for γ_R is

$$L = \frac{\widehat{\nu}_2}{\widehat{\gamma}_M F_{1-\alpha/2:m_1,m}}\left[\frac{\widehat{\nu}_1}{\widehat{\nu}_2} - F_{1-\alpha/2:m_1,\infty} + \frac{\widehat{\nu}_2 F_{1-\alpha/2:m_1,m_2}(F_{1-\alpha/2:m_1,\infty} - F_{1-\alpha/2:m_1,m_2})}{\widehat{\nu}_1}\right]$$

and

$$U = \frac{\widehat{\nu}_2}{\widehat{\gamma}_M F_{\alpha/2:m_1,m}}\left[\frac{\widehat{\nu}_1}{\widehat{\nu}_2} - F_{\alpha/2:m_1,\infty} + \frac{\widehat{\nu}_2 F_{\alpha/2:m_1,m_2}(F_{\alpha/2:m_1,\infty} - F_{\alpha/2:m_1,m_2})}{\widehat{\nu}_1}\right],$$

where $\widehat{\gamma}_M$ is defined in Equation (8.38),

$$m = \frac{\widehat{\gamma}_M^2 (sr)^2}{S_2^4/n_2 + (s-1)^2 S_4^4/n_4 + s^2(r-1)^2 S_5^4/n_5},$$

and $\widehat{\nu}_1$, $\widehat{\nu}_2$, m_1, and m_2 are defined in Equation (8.36).

Table 8.24. *GPQs for misclassification rates in model* (8.34).

Parameter	GPQ
$\mu_Y = \mu_P$	$\overline{y}_{****} - Z\sqrt{\max\left[\epsilon, \frac{n_1 s_1^2}{dosr W_1} + \frac{n_3 s_3^2}{dosr W_3} - \frac{n_4 s_4^2}{dosr W_4}\right]}$
$\gamma_P + \gamma_M$	$\frac{n_1 s_1^2}{osr W_1} + \frac{(o-1) n_2 s_2^2}{osr W_2} + \frac{n_3 s_3^2}{dor W_3} + \frac{(dos - do - s) n_4 s_4^2}{dosr W_4} + \frac{(r-1) n_5 s_5^2}{r W_5}$
γ_P	$\max\left[\epsilon, \frac{n_1 s_1^2}{osr W_1} + \frac{n_3 s_3^2}{dor W_3} - \frac{n_2 s_2^2}{osr W_2} - \frac{n_4 s_4^2}{dor W_4}\right]$

Table 8.25. 95% *confidence intervals for misclassification rates. (Bounds for* δ*,* δ_c*, and* β_c *have been multiplied by* 10^6*. The lower bound for* β *has been multiplied by* 10^{12}*.)*

Parameter	Lower bound	Upper bound
δ	1,332	65,700
δ_c	1,332	89,499
δ_{index}	0.35	12,639,781
β	5	41,320
β_c	41,094	463,818
β_{index}	0.049	0.464

For the particular example being considered, the computed 95% interval for γ_R with $\widehat{\nu}_1 = 0.5542, \widehat{\nu}_2 = 0.04647, m_1 = 4$ (truncated), $m_2 = 32$ (truncated), $m = 237$ (truncated), and $\widehat{\gamma}_M = 0.7560$ is from $L = 0.195$ to $U = 6.01$. By transforming the interval for γ_R, the 95% confidence interval for the SNR is from $L = \sqrt{2(0.195)} = 0.625$ to $U = \sqrt{2(6.01)} = 3.47$. The 95% GCI for γ_R formed by taking the ratio of the GPQs in Equations (8.37) and (8.39) is from $L = 0.254$ to $U = 7.78$. This transforms into a 95% confidence interval for SNR from $L = 0.713$ to $U = 3.94$.

8.5.4 Intervals for Misclassification Rates

Table 8.24 shows the GPQs required to compute confidence intervals for the misclassification rates. Table 8.25 shows the computed 95% intervals where $LSL = 304.7$, $USL = 311.1$, $\overline{y}_{****} = 307.9$, and $\epsilon = 0.001$.

8.5.5 Conclusions

The R&R plot in Figure 8.6 and the misclassification rates in Table 8.25 suggest this is not a very good measurement system. The residual plots in Figure 8.7 suggest the data are more skewed than a normal population.

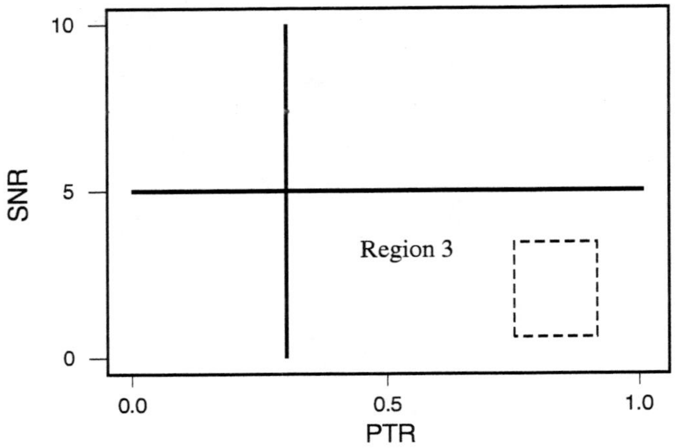

Figure 8.6. *R&R graph for semiconductor example.*

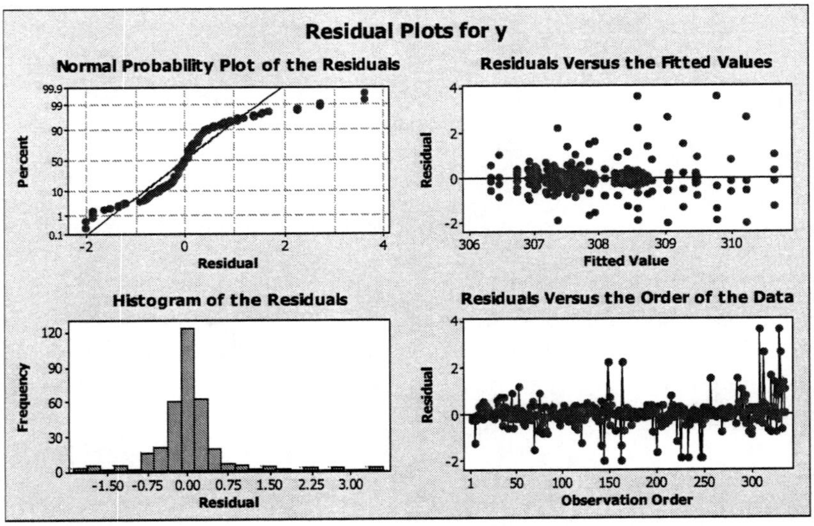

Figure 8.7. *Residual plots for the semiconductor study.*

8.6 Summary

In this chapter we presented general strategies for analyzing a gauge R&R study under any random or mixed experimental design. Three examples were provided to demonstrate these strategies. The strategies can be employed for both balanced and unbalanced designs.

Appendix A

The Analysis of Variance

The analysis of variance (ANOVA) is one of the most widely used techniques in applied statistics. It plays a central role in the analysis of gauge R&R studies. The objective of this appendix is to give an elementary description of the mechanics and underlying distribution theory of the ANOVA.

Suppose we have p different levels of a single factor that we wish to compare. These different factor levels are sometimes called treatments. Each treatment is replicated r times. The observed response from each of the p treatments is a random variable. The data usually appear as in Table A.1. An entry in this table (Y_{ij}) represents the jth observation taken under the ith treatment. In a gauge R&R study, the observations are measurements on the p parts. The model for the data in Table A.1 was presented in Equation (2.1) as

$$Y_{ij} = \mu_Y + P_i + E_{ij}, \qquad i = 1, \ldots, p, \quad j = 1, \ldots, r, \tag{A.1}$$

where μ_Y is a constant, P_i is a random variable that represents the parts, and E_{ij} is a random error term. The P_i are often called the treatment effects. We assume that P_i and E_{ij} are jointly independent normal random variables with means of zero and variances σ_P^2 and σ_E^2, respectively.

Model (A.1) is referred to as a random effects model because the treatments (parts) represent a random sample from a larger population of treatments. This is why the term P_i is a random variable. In the random effects model, one wants to extend the conclusions (which are based on the sample of treatments) to all treatments in the population.

In some statistical experiments, treatments are specifically chosen by the experimenter. In this situation one wishes to test hypotheses about the treatment means, and conclusions will apply only to the specific treatments considered in the analysis. The conclusions cannot be extended to similar treatments that were not explicitly considered. This is called the fixed effects model. The fixed effects one-factor model is

$$Y_{ij} = \mu_i + E_{ij}, \qquad i = 1, \ldots, p, \quad j = 1, \ldots, r, \tag{A.2}$$

Table A.1. *Data for a one-factor ANOVA.*

Treatment	Observations				Totals	Averages
1	Y_{11}	Y_{12}	$\dots$	Y_{1r}	Y_{1*}	$\overline{Y}_{1*}$
2	Y_{21}	Y_{22}	$\dots$	Y_{2r}	Y_{2*}	$\overline{Y}_{2*}$
$\vdots$	$\vdots$	$\vdots$	$\dots$	$\vdots$	$\vdots$	$\vdots$
p	Y_{p1}	Y_{p2}	$\dots$	Y_{pr}	Y_{p*}	$\overline{Y}_{p*}$
					Y_{**}	$\overline{Y}_{**}$

where μ_i is the mean of treatment i and E_{ij} is defined in the same manner as in model (A.1). Primary interest in the fixed model is to compare the μ_i.

Finally, models that contain both fixed and random effects are called mixed effects models. These models are studied in Chapter 6. This appendix concentrates on the random effects model.

We now present a simple development of the balanced one-factor random ANOVA introduced in Chapter 2. Notice that Table A.1 contains definitions of the totals and averages of the responses under each treatment and the grand total and grand average of all observations. The name "analysis of variance" is derived from a partitioning of total variability into its component parts. The total corrected sum of squares,

$$SS_T = \sum_{i=1}^{p}\sum_{j=1}^{r}(Y_{ij} - \overline{Y}_{**})^2,$$

is used as a measure of overall variability in the data. Intuitively, this is reasonable because if we were to divide SS_T by the appropriate degrees of freedom (in this case, $pr - 1$), this ratio would be the sample variance of the observations. Note that SS_T can be written as

$$\sum_{i=1}^{p}\sum_{j=1}^{r}(Y_{ij} - \overline{Y}_{**})^2 = \sum_{i=1}^{p}\sum_{j=1}^{r}[(\overline{Y}_{i*} - \overline{Y}_{**}) + (Y_{ij} - \overline{Y}_{i*})]^2. \tag{A.3}$$

Expanding the right-hand side of Equation (A.3) and simplifying gives

$$\sum_{i=1}^{p}\sum_{j=1}^{r}(Y_{ij} - \overline{Y}_{**})^2 = r\sum_{i=1}^{p}(\overline{Y}_{i*} - \overline{Y}_{**})^2 + \sum_{i=1}^{p}\sum_{j=1}^{r}(Y_{ij} - \overline{Y}_{i*})^2. \tag{A.4}$$

Equation (A.4) states that the total variability in the data, as measured by the total corrected sum of squares, can be partitioned into a sum of squares of the differences between the treatment averages and the grand average, plus a sum of squares of the differences of observations within treatments from the treatment average. Now, the difference between the observed treatment averages and the grand average is a measure of the differences between treatment means, and the differences of observations within a treatment from the treatment average is due only to random error. Thus, we may write Equation (A.4) symbolically as

$$SS_T = SS_{Treatments} + SS_E,$$

where $SS_{Treatments}$ is called the sum of squares between treatments, and SS_E is called the sum of squares due to error (i.e., within treatments). There are pr total observations. Thus, SS_T has $pr - 1$ degrees of freedom. There are p levels of the factor, so $SS_{Treatments}$ has $p - 1$ degrees of freedom. Finally, within any treatment there are r replicates providing $r - 1$ degrees of freedom for experimental error. Because there are p treatments, we have $p(r - 1)$ degrees of freedom for error. The quantities

$$S^2_{Treatments} = \frac{SS_{Treatments}}{p - 1}$$

and

$$S^2_E = \frac{SS_E}{p(r - 1)}$$

are called mean squares. (In Chapter 2 we represent $S^2_{Treatments}$ as S^2_P since "parts" is the treatment effect.) The expected values of these mean squares were given in Table 2.2, but for convenience they are repeated here:

$$E(S^2_{Treatments}) = \theta_P = \sigma^2_E + r\sigma^2_P$$

and

$$E(S^2_E) = \theta_E = \sigma^2_E. \tag{A.5}$$

Derivations of these expected mean squares are provided in [51].

The ANOVA is usually presented as a hypothesis testing procedure. In the random effects model, we are interested in testing a hypothesis that relates to the homogeneity of all of the treatments in the sampled population. Now if all treatments in the population are identical, then $\sigma^2_P = 0$. Therefore, the appropriate hypotheses are

$$H_0 : \sigma^2_P = 0, \tag{A.6}$$
$$H_a : \sigma^2_P > 0.$$

If the normality and independence assumptions made previously are true, then $SS_{Treatments}/\theta_P$ is a chi-squared random variable with $p-1$ degrees of freedom, SS_E/θ_E is a chi-squared random variable with $p(r-1)$ degrees of freedom, and these two chi-squared random variables are independent. Consequently, the ratio

$$\frac{S^2_{Treatments}}{S^2_E}\left(\frac{\theta_E}{\theta_P}\right)$$

is an F random variable with $p-1$ numerator degrees of freedom and $p(r-1)$ denominator degrees of freedom. From examining the expected mean squares in Equation (A.5), we note that if the null hypothesis is true and $\sigma^2_P = 0$, then $\theta_P = \theta_E$ so that

$$F_0 = \frac{S^2_{Treatments}}{S^2_E}$$

is an F random variable with $p-1$ numerator degrees of freedom and $p(r-1)$ denominator degrees of freedom. Thus, if $H_0 : \sigma^2_P = 0$ is true, observed values of F_0 will be consistent with those of an F-distribution. However, if H_0 is false, $\theta_P > \theta_E$ and F_0 will provide values greater than that expected from the F-distribution. The percentile $F_{1-\alpha:p-1,p(r-1)}$ is taken as the critical value where $1 - \alpha$ is the area to the left (see Section 1.7). Therefore, we reject the null hypothesis in (A.6) with level of significance α and conclude that $\sigma_P^2 > 0$ if $F_0 > F_{1-\alpha:p-1,p(r-1)}$.

In this book we focus on confidence intervals as opposed to hypothesis testing. One reason is because confidence intervals can be used to test hypotheses. For example, the hypotheses in (A.6) can be tested with a level of significance α using the lower bound of a $1-2\alpha$ two-sided confidence interval for σ_P^2 based on Equation (2.3). The null hypothesis is rejected if the lower bound is greater than zero. This will happen only if $F_0 > F_{1-\alpha:p-1,p(r-1)}$, which is the critical value discussed above.

The extension of ANOVA to more complicated treatment structure is relatively straightforward. Generally the total sum of squares is partitioned into components that are consistent with the underlying model for the experiment. These sums of squares are divided by their degrees of freedom to produce mean squares. In balanced random models, these mean squares are distributed as independent chi-squared random variables. Ratios of the mean squares divided by their expected values are distributed as F random variables. In some experimental designs, construction of relevant test statistics for the hypotheses of interest requires careful inspection of the expected means squares. Details for many different designs and ANOVA models are given in [51].

Appendix B

MLS and GCI Methods

In this appendix we describe the rationale for the MLS and GCI approaches. We do this in the context of a balanced random model with the ANOVA shown in Table B.1. Given the standard normality and independence assumptions made in this book, $n_q S_q^2/\theta_q$ are jointly independent chi-squared random variables with n_q degrees of freedom.

Table B.1. *General ANOVA for a balanced random model.*

Source of variation	Degrees of freedom	Mean square	Expected mean square
Factor 1	n_1	S_1^2	θ_1
Factor 2	n_2	S_2^2	θ_2
$\vdots$	$\vdots$	$\vdots$	$\vdots$
Replicates	n_Q	S_Q^2	θ_Q

B.1 MLS Method

The term "modified large-sample" was introduced by Graybill and Wang [25] in their paper proposing a confidence interval for nonnegative linear combinations of variance components. To apply this method, one starts with an approximate large-sample confidence interval. The interval is then modified to make it exact under certain parameter conditions. To demonstrate, consider the problem of constructing a confidence interval for $\gamma = \theta_1 + \theta_2$, where θ_1 and θ_2 are defined in Table B.1. The uniformly minimum variance unbiased estimator for γ is $S_1^2 + S_2^2$ and this estimator has variance $2\theta_1^2/n_1 + 2\theta_2^2/n_2$. Thus, the $100(1-\alpha)\%$ large-sample confidence interval for γ is

$$\begin{aligned} L &= S_1^2 + S_2^2 - Z_{1-\alpha/2}\sqrt{2\theta_1^2/n_1 + 2\theta_2^2/n_2} \\ &= S_1^2 + S_2^2 - \sqrt{(2Z_{1-\alpha/2}^2/n_1)\theta_1^2 + (2Z_{1-\alpha/2}^2/n_2)\theta_2^2} \end{aligned}$$

and

$$U = S_1^2 + S_2^2 + Z_{1-\alpha/2}\sqrt{2\theta_1^2/n_1 + 2\theta_2^2/n_2}$$
$$= S_1^2 + S_2^2 + \sqrt{(2Z_{1-\alpha/2}^2/n_1)\theta_1^2 + (2Z_{1-\alpha/2}^2/n_2)\theta_2^2},$$

where $Z_{1-\alpha/2}$ is the percentile from a standard normal distribution with area $1-\alpha/2$ to the left. We now replace θ_1 and θ_2 with the estimators S_1^2 and S_2^2, and replace the constants $2Z_{1-\alpha/2}^2/n_1$ and $2Z_{1-\alpha/2}^2/n_2$ with general constants to yield the interval

$$L = S_1^2 + S_2^2 - \sqrt{G_1^2 S_1^4 + G_2^2 S_2^4}$$

and

$$U = S_1^2 + S_2^2 + \sqrt{H_1^2 S_1^4 + H_2^2 S_2^4}. \tag{B.1}$$

The constants G_1, G_2, H_1, and H_2 are now selected so that the interval in Equation (B.1) has an exact confidence coefficient of $100(1-\alpha)\%$ when $\theta_1 = 0$ or $\theta_2 = 0$. When $\theta_1 = 0$, it follows that $S_1^2 = 0$ with probability one, and

$$L = S_2^2 - \sqrt{G_2^2 S_2^4}$$
$$= S_2^2 - G_2 S_2^2$$
$$= S_2^2(1 - G_2)$$

and

$$U = S_2^2 + \sqrt{H_2^2 S_2^4}$$
$$= S_2^2 + H_2 S_2^2$$
$$= S_2^2(1 + H_2). \tag{B.2}$$

When $\theta_1 = 0$, an exact $100(1-\alpha)\%$ confidence interval for $\gamma = \theta_2$ is

$$L = \frac{n_2 S_2^2}{\chi_{1-\alpha/2:n_2}^2}$$

and

$$U = \frac{n_2 S_2^2}{\chi_{\alpha/2:n_2}^2}. \tag{B.3}$$

Now by selecting G_2 and H_2 so that Equation (B.2) reduces to Equation (B.3), the interval in Equation (B.2) will have an exact confidence coefficient of $100(1-\alpha)\%$. This requires

$$G_2 = 1 - \frac{n_2}{\chi_{1-\alpha/2:n_2}^2}$$

and

$$H_2 = \frac{n_2}{\chi_{\alpha/2:n_2}^2} - 1.$$

From Equation (1.16),

$$\frac{\chi^2_{\alpha:df}}{df} = F_{\alpha:df,\infty}$$

and

$$\frac{1}{F_{\alpha:df1,df2}} = F_{1-\alpha:df2,df1}.$$

Thus,

$$G_2 = 1 - F_{\alpha/2:\infty,n_2}$$

and

$$H_2 = F_{1-\alpha/2:\infty,n_2} - 1.$$

Similarly, if $\theta_2 = 0$, interval (B.1) is exact if $G_1 = 1 - F_{\alpha/2:\infty,n_1}$ and $H_1 = F_{1-\alpha/2:\infty,n_1} - 1$. Note that these constants also imply that the interval in Equation (B.1) is exact when $n_1 \to \infty$ and n_2 is fixed or when $n_2 \to \infty$ and n_1 is fixed.

This general approach is followed for all the MLS intervals, although the conditions for exactness can vary. For example, the MLS interval proposed by Ting et al. [64] for $\theta_1 - \theta_2$ is exact when $\theta_1 = 0$, or $\theta_2 = 0$, or $\theta_1 = \theta_2$. By making the MLS intervals exact for limiting conditions, it is hoped that the confidence coefficient will be close to the stated level for other configurations of parameters. Empirical evidence suggests that the MLS intervals generally provide confidence coefficients at least as great as the stated level.

B.2 GCIs

GCIs were introduced by Weerahandi [71]. The concept of GCI is a natural extension of generalized p-values and generalized test variables introduced by Tsui and Weerahandi [65]. It relies on the idea of a GPQ, an extension of the standard notion of a pivotal quantity. The method can occasionally lead to closed-form solutions, but as a rule the confidence bounds are approximated using Monte Carlo methods.

Suppose we wish to construct a GCI for a scalar parameter γ. Let S represent a random variable whose distribution depends on γ and a nuisance parameter δ (possibly a vector parameter). A GPQ, say, $T = T(S; s, \gamma, \delta)$, is a function of S, the realized value of S (denoted by s), γ, and δ, that satisfies the following two conditions:

1. For fixed s, the distribution of T does not depend on any unknown parameters.
2. The observed value of T, $T_{obs} = T(s; s, \gamma, \delta)$, is equal to γ.

Let C_α be a region satisfying

$$\Pr[T \in C_\alpha] = 1 - \alpha.$$

A $100(1-\alpha)\%$ generalized confidence region for γ is the set

$$\{\gamma : T(s; s, \gamma, \delta) \in C_\alpha\}.$$

For instance, one may take C_α to be the interval $[T_{\alpha/2}, T_{1-\alpha/2}]$, where $T_{\alpha/2}$ is the $\alpha/2$ percentile of the distribution of T and $T_{1-\alpha/2}$ is the $1-\alpha/2$ percentile of the distribution of T. These percentiles can be obtained in closed form only rarely, but they can be estimated using Monte Carlo methods.

Note that property 1 guarantees that C_α does not depend on the parameters γ and δ, and property 2 guarantees that the generalized confidence region can be constructed using only observed data. Weerahandi [72, Chapter 6] provides more details on generalized confidence regions.

One method for determining an appropriate GPQ is as follows. Consider the ANOVA in Table B.1 and assume the parameter of interest is $\gamma = \theta_1 + \theta_2$. Note that $n_1 S_1^2/\theta_1$ and $n_2 S_2^2/\theta_2$ are jointly independent chi-squared random variables with n_1 and n_2 degrees of freedom, respectively. Thus we can write

$$\frac{n_1 S_1^2}{\theta_1} = W_1$$

and

$$\frac{n_2 S_2^2}{\theta_2} = W_2, \tag{B.4}$$

where W_1 and W_2 and independent chi-squared random variables with n_1 and n_2 degrees of freedom, respectively. We now invert the pivotal quantities in Equation (B.4) and write

$$\theta_1 = \frac{n_1 S_1^2}{W_1}$$

and

$$\theta_2 = \frac{n_2 S_2^2}{W_2}.$$

Next replace the random variables S_1^2 and S_2^2 with their realized sample values s_1^2 and s_2^2 to form the quantities

$$T_1 = \frac{n_1 s_1^2}{W_1}$$

and

$$T_2 = \frac{n_2 s_2^2}{W_2}.$$

The random variables T_1 and T_2 are the GPQs for the parameters θ_1 and θ_2, respectively. To form a GPQ for $\gamma = \theta_1 + \theta_2$, we simply sum T_1 and T_2. That is, the GPQ for γ is

$$\frac{n_1 s_1^2}{W_1} + \frac{n_2 s_2^2}{W_2}.$$

This is the process we use throughout the book for defining GPQs.

Iyer and Patterson [34] formalized this approach and provide theoretical rationale for the process. They also provide several applications of this approach to a variety of problems.

Appendix C

Tables of *F*-Values

The following tables can be used in the confidence interval formulas presented in the book. For an explanation of the notation used in these tables, please refer to Section 1.7.

Table C.1. *F-values for 80% two-sided intervals ($F_{0.10:r,c}$ and $F_{0.90:r,c}$).*

r	c = 1	2	3	4	5	6	7	8	9
1	0.0251	0.0202	0.0187	0.0179	0.0175	0.0172	0.0170	0.0168	0.0167
	39.864	8.5263	5.5383	4.5448	4.0604	3.7760	3.5894	3.4579	3.3603
2	0.1173	0.1111	0.1091	0.1082	0.1076	0.1072	0.1070	0.1068	0.1066
	49.500	9.0000	5.4624	4.3246	3.7797	3.4633	3.2574	3.1131	3.0065
3	0.1806	0.1831	0.1855	0.1872	0.1884	0.1892	0.1899	0.1904	0.1908
	53.593	9.1618	5.3908	4.1909	3.6195	3.2888	3.0741	2.9238	2.8129
4	0.2200	0.2312	0.2386	0.2435	0.2469	0.2494	0.2513	0.2528	0.2541
	55.833	9.2434	5.3426	4.1072	3.5202	3.1808	2.9605	2.8064	2.6927
5	0.2463	0.2646	0.2763	0.2841	0.2896	0.2937	0.2969	0.2995	0.3015
	57.240	9.2926	5.3092	4.0506	3.4530	3.1075	2.8833	2.7264	2.6106
6	0.2648	0.2887	0.3041	0.3144	0.3218	0.3274	0.3317	0.3352	0.3381
	58.204	9.3255	5.2847	4.0097	3.4045	3.0546	2.8274	2.6683	2.5509
7	0.2786	0.3070	0.3253	0.3378	0.3468	0.3537	0.3591	0.3634	0.3670
	58.906	9.3491	5.2662	3.9790	3.3679	3.0145	2.7849	2.6241	2.5053
8	0.2892	0.3212	0.3420	0.3563	0.3668	0.3748	0.3811	0.3862	0.3904
	59.439	9.3668	5.2517	3.9549	3.3393	2.9830	2.7516	2.5893	2.4694
9	0.2976	0.3326	0.3555	0.3714	0.3831	0.3920	0.3992	0.4050	0.4098
	59.858	9.3805	5.2400	3.9357	3.3163	2.9577	2.7247	2.5612	2.4403
10	0.3044	0.3419	0.3666	0.3838	0.3966	0.4064	0.4143	0.4207	0.4260
	60.195	9.3916	5.2304	3.9199	3.2974	2.9369	2.7025	2.5380	2.4163
12	0.3148	0.3563	0.3838	0.4032	0.4177	0.4290	0.4381	0.4455	0.4518
	60.705	9.4081	5.2156	3.8955	3.2682	2.9047	2.6681	2.5020	2.3789
15	0.3254	0.3710	0.4016	0.4235	0.4399	0.4529	0.4634	0.4720	0.4793
	61.220	9.4247	5.2003	3.8704	3.2380	2.8712	2.6322	2.4642	2.3396
20	0.3362	0.3862	0.4202	0.4447	0.4633	0.4782	0.4903	0.5004	0.5089
	61.740	9.4413	5.1845	3.8443	3.2067	2.8363	2.5947	2.4246	2.2983
25	0.3427	0.3955	0.4316	0.4578	0.4780	0.4941	0.5073	0.5183	0.5278
	62.055	9.4513	5.1747	3.8283	3.1873	2.8147	2.5714	2.3999	2.2725
30	0.3471	0.4018	0.4394	0.4668	0.4880	0.5050	0.5190	0.5308	0.5408
	62.265	9.4579	5.1681	3.8174	3.1741	2.8000	2.5555	2.3830	2.2547
60	0.3583	0.4178	0.4593	0.4900	0.5140	0.5334	0.5496	0.5634	0.5754
	62.794	9.4746	5.1512	3.7896	3.1402	2.7620	2.5142	2.3391	2.2085
120	0.3639	0.4260	0.4695	0.5019	0.5275	0.5483	0.5658	0.5807	0.5937
	63.061	9.4829	5.1425	3.7753	3.1228	2.7423	2.4928	2.3162	2.1843
∞	0.3696	0.4343	0.4799	0.5142	0.5413	0.5637	0.5825	0.5987	0.6129
	63.328	9.4912	5.1337	3.7607	3.1050	2.7222	2.4708	2.2926	2.1592

Table C.1 *continued.*

r	c								
	10	12	15	20	25	30	60	120	∞
1	0.0166	0.0165	0.0163	0.0162	0.0161	0.0161	0.0159	0.0159	0.0158
	3.2850	3.1765	3.0732	2.9747	2.9177	2.8807	2.7911	2.7478	2.7055
2	0.1065	0.1063	0.1061	0.1059	0.1058	0.1057	0.1055	0.1055	0.1054
	2.9245	2.8068	2.6952	2.5893	2.5283	2.4887	2.3933	2.3473	2.3026
3	0.1912	0.1917	0.1923	0.1929	0.1932	0.1935	0.1941	0.1945	0.1948
	2.7277	2.6055	2.4898	2.3801	2.3170	2.2761	2.1774	2.1300	2.0838
4	0.2551	0.2567	0.2584	0.2601	0.2612	0.2620	0.2639	0.2649	0.2659
	2.6053	2.4801	2.3614	2.2489	2.1842	2.1422	2.0410	1.9923	1.9449
5	0.3033	0.3060	0.3088	0.3119	0.3137	0.3151	0.3184	0.3202	0.3221
	2.5216	2.3940	2.2730	2.1582	2.0922	2.0492	1.9457	1.8959	1.8473
6	0.3405	0.3443	0.3483	0.3526	0.3553	0.3571	0.3621	0.3647	0.3674
	2.4606	2.3310	2.2081	2.0913	2.0241	1.9803	1.8747	1.8238	1.7741
7	0.3700	0.3748	0.3799	0.3854	0.3889	0.3913	0.3977	0.4012	0.4047
	2.4140	2.2828	2.1582	2.0397	1.9714	1.9269	1.8194	1.7675	1.7167
8	0.3940	0.3997	0.4058	0.4124	0.4167	0.4196	0.4275	0.4317	0.4362
	2.3771	2.2446	2.1185	1.9985	1.9292	1.8841	1.7748	1.7220	1.6702
9	0.4139	0.4204	0.4274	0.4351	0.4401	0.4435	0.4528	0.4578	0.4631
	2.3473	2.2135	2.0862	1.9649	1.8947	1.8490	1.7380	1.6842	1.6315
10	0.4306	0.4378	0.4457	0.4544	0.4600	0.4639	0.4746	0.4804	0.4865
	2.3226	2.1878	2.0593	1.9367	1.8658	1.8195	1.7070	1.6524	1.5987
12	0.4571	0.4657	0.4751	0.4855	0.4923	0.4971	0.5103	0.5175	0.5253
	2.2841	2.1474	2.0171	1.8924	1.8200	1.7727	1.6574	1.6012	1.5458
15	0.4856	0.4958	0.5070	0.5197	0.5280	0.5340	0.5504	0.5597	0.5698
	2.2435	2.1049	1.9722	1.8449	1.7708	1.7223	1.6034	1.5450	1.4871
20	0.5163	0.5284	0.5420	0.5575	0.5678	0.5753	0.5964	0.6085	0.6221
	2.2007	2.0597	1.9243	1.7938	1.7175	1.6673	1.5435	1.4821	1.4206
25	0.5360	0.5494	0.5647	0.5822	0.5941	0.6028	0.6276	0.6422	0.6589
	2.1739	2.0312	1.8939	1.7611	1.6831	1.6316	1.5039	1.4399	1.3753
30	0.5496	0.5641	0.5806	0.5998	0.6129	0.6225	0.6504	0.6672	0.6866
	2.1554	2.0115	1.8728	1.7382	1.6589	1.6065	1.4755	1.4094	1.3419
60	0.5858	0.6033	0.6237	0.6479	0.6649	0.6777	0.7167	0.7421	0.7743
	2.1072	1.9597	1.8168	1.6768	1.5934	1.5376	1.3952	1.3203	1.2400
120	0.6052	0.6245	0.6472	0.6747	0.6945	0.7095	0.7574	0.7908	0.8385
	2.0818	1.9323	1.7867	1.6433	1.5570	1.4989	1.3476	1.2646	1.1686
∞	0.6255	0.6469	0.6724	0.7039	0.7271	0.7452	0.8065	0.8557	1.0000
	2.0554	1.9036	1.7551	1.6074	1.5176	1.4564	1.2915	1.1926	1.0000

Table C.2. *F-values for 90% two-sided intervals ($F_{0.05:r,c}$ and $F_{0.95:r,c}$).*

	c								
r	1	2	3	4	5	6	7	8	9
1	0.0062 161.4	0.0050 18.513	0.0046 10.128	0.0045 7.7086	0.0043 6.6079	0.0043 5.9874	0.0042 5.5915	0.0042 5.3177	0.0042 5.1174
2	0.0540 199.5	0.0526 19.000	0.0522 9.5521	0.0520 6.9443	0.0518 5.7861	0.0517 5.1433	0.0517 4.7374	0.0516 4.4590	0.0516 4.2565
3	0.0987 215.7	0.1047 19.164	0.1078 9.2766	0.1097 6.5914	0.1109 5.4095	0.1118 4.7571	0.1125 4.3468	0.1131 4.0662	0.1135 3.8625
4	0.1297 224.6	0.1440 19.247	0.1517 9.1172	0.1565 6.3882	0.1598 5.1922	0.1623 4.5337	0.1641 4.1203	0.1655 3.8379	0.1667 3.6331
5	0.1513 230.2	0.1728 19.296	0.1849 9.0135	0.1926 6.2561	0.1980 5.0503	0.2020 4.3874	0.2051 3.9715	0.2075 3.6875	0.2095 3.4817
6	0.1670 234.0	0.1944 19.330	0.2102 8.9406	0.2206 6.1631	0.2279 4.9503	0.2334 4.2839	0.2377 3.8660	0.2411 3.5806	0.2440 3.3738
7	0.1788 236.8	0.2111 19.353	0.2301 8.8867	0.2427 6.0942	0.2518 4.8759	0.2587 4.2067	0.2641 3.7870	0.2684 3.5005	0.2720 3.2927
8	0.1881 238.9	0.2243 19.371	0.2459 8.8452	0.2606 6.0410	0.2712 4.8183	0.2793 4.1468	0.2857 3.7257	0.2909 3.4381	0.2951 3.2296
9	0.1954 240.5	0.2349 19.385	0.2589 8.8123	0.2752 5.9988	0.2872 4.7725	0.2964 4.0990	0.3037 3.6767	0.3096 3.3881	0.3146 3.1789
10	0.2014 241.9	0.2437 19.396	0.2697 8.7855	0.2875 5.9644	0.3007 4.7351	0.3108 4.0600	0.3189 3.6365	0.3256 3.3472	0.3311 3.1373
12	0.2106 243.9	0.2574 19.413	0.2865 8.7446	0.3068 5.9117	0.3220 4.6777	0.3338 3.9999	0.3432 3.5747	0.3511 3.2839	0.3576 3.0729
15	0.2201 245.9	0.2716 19.429	0.3042 8.7029	0.3273 5.8578	0.3447 4.6188	0.3584 3.9381	0.3695 3.5107	0.3787 3.2184	0.3865 3.0061
20	0.2298 248.0	0.2863 19.446	0.3227 8.6602	0.3489 5.8025	0.3689 4.5581	0.3848 3.8742	0.3978 3.4445	0.4087 3.1503	0.4179 2.9365
25	0.2358 249.3	0.2954 19.456	0.3343 8.6341	0.3625 5.7687	0.3842 4.5209	0.4015 3.8348	0.4158 3.4036	0.4279 3.1081	0.4382 2.8932
30	0.2398 250.1	0.3016 19.462	0.3422 8.6166	0.3718 5.7459	0.3947 4.4957	0.4131 3.8082	0.4284 3.3758	0.4413 3.0794	0.4523 2.8637
60	0.2499 252.2	0.3174 19.479	0.3626 8.5720	0.3960 5.6877	0.4222 4.4314	0.4436 3.7398	0.4616 3.3043	0.4769 3.0053	0.4902 2.7872
120	0.2551 253.3	0.3255 19.487	0.3731 8.5494	0.4086 5.6581	0.4367 4.3985	0.4598 3.7047	0.4792 3.2674	0.4959 2.9669	0.5105 2.7475
∞	0.2603 254.3	0.3338 19.496	0.3839 8.5264	0.4216 5.6281	0.4517 4.3650	0.4765 3.6689	0.4976 3.2298	0.5159 2.9276	0.5319 2.7067

Table C.2 *continued.*

r	c: 10	12	15	20	25	30	60	120	∞
1	0.0041 4.9646	0.0041 4.7472	0.0041 4.5431	0.0040 4.3512	0.0040 4.2417	0.0040 4.1709	0.0040 4.0012	0.0039 3.9201	0.0039 3.8415
2	0.0516 4.1028	0.0515 3.8853	0.0515 3.6823	0.0514 3.4928	0.0514 3.3852	0.0514 3.3158	0.0513 3.1504	0.0513 3.0718	0.0513 2.9957
3	0.1138 3.7083	0.1144 3.4903	0.1149 3.2874	0.1155 3.0984	0.1158 2.9912	0.1161 2.9223	0.1167 2.7581	0.1170 2.6802	0.1173 2.6049
4	0.1677 3.4780	0.1692 3.2592	0.1707 3.0556	0.1723 2.8661	0.1733 2.7587	0.1740 2.6896	0.1758 2.5252	0.1767 2.4472	0.1777 2.3719
5	0.2112 3.3258	0.2138 3.1059	0.2165 2.9013	0.2194 2.7109	0.2212 2.6030	0.2224 2.5336	0.2257 2.3683	0.2274 2.2899	0.2291 2.2141
6	0.2463 3.2172	0.2500 2.9961	0.2539 2.7905	0.2581 2.5990	0.2608 2.4904	0.2626 2.4205	0.2674 2.2541	0.2699 2.1750	0.2726 2.0986
7	0.2750 3.1355	0.2797 2.9134	0.2848 2.7066	0.2903 2.5140	0.2938 2.4047	0.2962 2.3343	0.3026 2.1665	0.3060 2.0868	0.3096 2.0096
8	0.2988 3.0717	0.3045 2.8486	0.3107 2.6408	0.3174 2.4471	0.3217 2.3371	0.3247 2.2662	0.3327 2.0970	0.3370 2.0164	0.3416 1.9384
9	0.3187 3.0204	0.3254 2.7964	0.3327 2.5876	0.3405 2.3928	0.3456 2.2821	0.3492 2.2107	0.3588 2.0401	0.3640 1.9588	0.3695 1.8799
10	0.3358 2.9782	0.3433 2.7534	0.3515 2.5437	0.3605 2.3479	0.3663 2.2365	0.3704 2.1646	0.3815 1.9926	0.3876 1.9105	0.3940 1.8307
12	0.3632 2.9130	0.3722 2.6866	0.3821 2.4753	0.3931 2.2776	0.4004 2.1649	0.4055 2.0921	0.4194 1.9174	0.4272 1.8337	0.4355 1.7522
15	0.3931 2.8450	0.4040 2.6169	0.4161 2.4034	0.4296 2.2033	0.4386 2.0889	0.4451 2.0148	0.4629 1.8364	0.4730 1.7505	0.4841 1.6664
20	0.4259 2.7740	0.4391 2.5436	0.4539 2.3275	0.4708 2.1242	0.4822 2.0075	0.4904 1.9317	0.5138 1.7480	0.5273 1.6587	0.5425 1.5705
25	0.4471 2.7298	0.4619 2.4977	0.4787 2.2797	0.4981 2.0739	0.5114 1.9554	0.5211 1.8782	0.5489 1.6902	0.5655 1.5980	0.5845 1.5061
30	0.4620 2.6996	0.4780 2.4663	0.4963 2.2468	0.5177 2.0391	0.5324 1.9192	0.5432 1.8409	0.5749 1.6491	0.5940 1.5543	0.6164 1.4591
60	0.5019 2.6211	0.5215 2.3842	0.5445 2.1601	0.5721 1.9464	0.5916 1.8217	0.6064 1.7396	0.6518 1.5343	0.6815 1.4290	0.7198 1.3180
120	0.5234 2.5801	0.5453 2.3410	0.5713 2.1141	0.6029 1.8963	0.6258 1.7684	0.6434 1.6835	0.6998 1.4673	0.7397 1.3519	0.7975 1.2214
∞	0.5462 2.5379	0.5707 2.2962	0.6001 2.0658	0.6367 1.8432	0.6640 1.7110	0.6854 1.6223	0.7587 1.3893	0.8187 1.2539	1.0000 1.0000

Table C.3. *F-values for 95% two-sided intervals ($F_{0.025:r,c}$ and $F_{0.975:r,c}$).*

r	c = 1	2	3	4	5	6	7	8	9
1	0.0015	0.0013	0.0012	0.0011	0.0011	0.0011	0.0011	0.0010	0.0010
	647.8	38.5063	17.4434	12.2179	10.0070	8.8131	8.0727	7.5709	7.2093
2	0.0260	0.0256	0.0255	0.0255	0.0254	0.0254	0.0254	0.0254	0.0254
	799.5	39.0000	16.0441	10.6491	8.4336	7.2599	6.5415	6.0595	5.7147
3	0.0573	0.0623	0.0648	0.0662	0.0672	0.0679	0.0684	0.0688	0.0691
	864.2	39.1655	15.4392	9.9792	7.7636	6.5988	5.8898	5.4160	5.0781
4	0.0818	0.0939	0.1002	0.1041	0.1068	0.1087	0.1102	0.1114	0.1123
	899.6	39.2484	15.1010	9.6045	7.3879	6.2272	5.5226	5.0526	4.7181
5	0.0999	0.1186	0.1288	0.1354	0.1399	0.1433	0.1459	0.1480	0.1497
	921.8	39.2982	14.8848	9.3645	7.1464	5.9876	5.2852	4.8173	4.4844
6	0.1135	0.1377	0.1515	0.1606	0.1670	0.1718	0.1756	0.1786	0.1810
	937.1	39.3315	14.7347	9.1973	6.9777	5.8198	5.1186	4.6517	4.3197
7	0.1239	0.1529	0.1698	0.1811	0.1892	0.1954	0.2002	0.2041	0.2073
	948.2	39.3552	14.6244	9.0741	6.8531	5.6955	4.9949	4.5286	4.1970
8	0.1321	0.1650	0.1846	0.1979	0.2076	0.2150	0.2208	0.2256	0.2295
	956.7	39.3730	14.5399	8.9796	6.7572	5.5996	4.8993	4.4333	4.1020
9	0.1387	0.1750	0.1969	0.2120	0.2230	0.2315	0.2383	0.2438	0.2484
	963.3	39.3869	14.4731	8.9047	6.6811	5.5234	4.8232	4.3572	4.0260
10	0.1442	0.1833	0.2072	0.2238	0.2361	0.2456	0.2532	0.2594	0.2646
	968.6	39.3980	14.4189	8.8439	6.6192	5.4613	4.7611	4.2951	3.9639
12	0.1526	0.1962	0.2235	0.2426	0.2570	0.2682	0.2773	0.2848	0.2910
	976.7	39.4146	14.3366	8.7512	6.5245	5.3662	4.6658	4.1997	3.8682
15	0.1613	0.2099	0.2408	0.2629	0.2796	0.2929	0.3036	0.3126	0.3202
	984.9	39.4313	14.2527	8.6565	6.4277	5.2687	4.5678	4.1012	3.7694
20	0.1703	0.2242	0.2592	0.2845	0.3040	0.3197	0.3325	0.3433	0.3525
	993.1	39.4479	14.1674	8.5599	6.3286	5.1684	4.4667	3.9995	3.6669
25	0.1759	0.2330	0.2707	0.2982	0.3196	0.3369	0.3511	0.3632	0.3736
	998.1	39.4579	14.1155	8.5010	6.2679	5.1069	4.4045	3.9367	3.6035
30	0.1796	0.2391	0.2786	0.3077	0.3304	0.3488	0.3642	0.3772	0.3884
	1001.4	39.4646	14.0805	8.4613	6.2269	5.0652	4.3624	3.8940	3.5604
60	0.1892	0.2548	0.2992	0.3325	0.3589	0.3806	0.3989	0.4147	0.4284
	1009.8	39.4812	13.9921	8.3604	6.1225	4.9589	4.2544	3.7844	3.4493
120	0.1941	0.2628	0.3099	0.3455	0.3740	0.3976	0.4176	0.4349	0.4501
	1014.0	39.4896	13.9473	8.3092	6.0693	4.9044	4.1989	3.7279	3.3918
∞	0.1990	0.2711	0.3209	0.3590	0.3896	0.4152	0.4372	0.4562	0.4731
	1018.3	39.4979	13.9021	8.2573	6.0153	4.8491	4.1423	3.6702	3.3329

Table C.3 *continued.*

r	c: 10	12	15	20	25	30	60	120	∞
1	0.0010 6.9367	0.0010 6.5538	0.0010 6.1995	0.0010 5.8715	0.0010 5.6864	0.0010 5.5675	0.0010 5.2856	0.0010 5.1523	0.0010 5.0239
2	0.0254 5.4564	0.0254 5.0959	0.0254 4.7651	0.0253 4.4613	0.0253 4.2909	0.0253 4.1821	0.0253 3.9253	0.0253 3.8046	0.0253 3.6889
3	0.0694 4.8256	0.0698 4.4742	0.0702 4.1528	0.0706 3.8587	0.0708 3.6943	0.0710 3.5894	0.0715 3.3425	0.0717 3.2269	0.0719 3.1161
4	0.1131 4.4683	0.1143 4.1212	0.1155 3.8043	0.1168 3.5147	0.1176 3.3530	0.1182 3.2499	0.1196 3.0077	0.1203 2.8943	0.1211 2.7858
5	0.1511 4.2361	0.1533 3.8911	0.1556 3.5764	0.1580 3.2891	0.1595 3.1287	0.1606 3.0265	0.1633 2.7863	0.1648 2.6740	0.1662 2.5665
6	0.1831 4.0721	0.1864 3.7283	0.1898 3.4147	0.1935 3.1283	0.1958 2.9686	0.1974 2.8667	0.2017 2.6274	0.2039 2.5154	0.2062 2.4082
7	0.2100 3.9498	0.2143 3.6065	0.2189 3.2934	0.2239 3.0074	0.2270 2.8478	0.2292 2.7460	0.2351 2.5068	0.2382 2.3948	0.2414 2.2875
8	0.2328 3.8549	0.2381 3.5118	0.2438 3.1987	0.2500 2.9128	0.2540 2.7531	0.2568 2.6513	0.2642 2.4117	0.2682 2.2994	0.2725 2.1918
9	0.2523 3.7790	0.2585 3.4358	0.2653 3.1227	0.2727 2.8365	0.2775 2.6766	0.2809 2.5746	0.2899 2.3344	0.2948 2.2217	0.3000 2.1136
10	0.2690 3.7168	0.2762 3.3736	0.2840 3.0602	0.2925 2.7737	0.2981 2.6135	0.3020 2.5112	0.3127 2.2702	0.3185 2.1570	0.3247 2.0483
12	0.2964 3.6209	0.3051 3.2773	0.3147 2.9633	0.3254 2.6758	0.3325 2.5149	0.3375 2.4120	0.3512 2.1692	0.3588 2.0548	0.3670 1.9447
15	0.3268 3.5217	0.3375 3.1772	0.3494 2.8621	0.3629 2.5731	0.3718 2.4110	0.3783 2.3072	0.3962 2.0613	0.4063 1.9450	0.4175 1.8326
20	0.3605 3.4185	0.3737 3.0728	0.3886 2.7559	0.4058 2.4645	0.4174 2.3005	0.4258 2.1952	0.4498 1.9445	0.4638 1.8249	0.4795 1.7085
25	0.3826 3.3546	0.3976 3.0077	0.4148 2.6894	0.4347 2.3959	0.4484 2.2303	0.4584 2.1237	0.4874 1.8687	0.5048 1.7462	0.5248 1.6259
30	0.3982 3.3110	0.4146 2.9633	0.4334 2.6437	0.4555 2.3486	0.4709 2.1816	0.4822 2.0739	0.5155 1.8152	0.5358 1.6899	0.5597 1.5660
60	0.4405 3.1984	0.4610 2.8478	0.4851 2.5242	0.5143 2.2234	0.5351 2.0516	0.5509 1.9400	0.6000 1.6668	0.6325 1.5299	0.6747 1.3883
120	0.4636 3.1399	0.4867 2.7874	0.5141 2.4611	0.5480 2.1562	0.5727 1.9811	0.5917 1.8664	0.6536 1.5810	0.6980 1.4327	0.7631 1.2684
∞	0.4882 3.0798	0.5142 2.7249	0.5457 2.3953	0.5853 2.0853	0.6151 1.9055	0.6386 1.7867	0.7203 1.4821	0.7884 1.3104	1.0000 1.0000

Table C.4. *F-values for 99% two-sided intervals* ($F_{0.005:r,c}$ *and* $F_{0.995:r,c}$).

r	c = 1	2	3	4	5	6	7	8	9
1	6.2E-05 16211	5.0E-05 198.5	4.6E-05 55.5520	4.4E-05 31.3328	4.3E-05 22.7848	4.3E-05 18.6350	4.2E-05 16.2356	4.2E-05 14.6882	4.2E-05 13.6136
2	0.005038 20000	0.005025 199.0	0.005021 49.7993	0.005019 26.2843	0.005018 18.3138	0.005017 14.5441	0.005016 12.4040	0.005016 11.0424	0.005015 10.1067
3	0.0180 21615	0.0201 199.2	0.0211 47.4672	0.0216 24.2591	0.0220 16.5298	0.0223 12.9166	0.0225 10.8824	0.0227 9.5965	0.0228 8.7171
4	0.0319 22500	0.0380 199.2	0.0412 46.1946	0.0432 23.1545	0.0445 15.5561	0.0455 12.0275	0.0462 10.0505	0.0468 8.8051	0.0473 7.9559
5	0.0439 23056	0.0546 199.3	0.0605 45.3916	0.0643 22.4564	0.0669 14.9396	0.0689 11.4637	0.0704 9.5221	0.0716 8.3018	0.0726 7.4712
6	0.0537 23437	0.0688 199.3	0.0774 44.8385	0.0831 21.9746	0.0872 14.5133	0.0903 11.0730	0.0927 9.1553	0.0946 7.9520	0.0962 7.1339
7	0.0616 23715	0.0806 199.4	0.0919 44.4341	0.0995 21.6217	0.1050 14.2004	0.1092 10.7859	0.1125 8.8854	0.1152 7.6941	0.1175 6.8849
8	0.0681 23925	0.0906 199.4	0.1042 44.1256	0.1136 21.3520	0.1205 13.9610	0.1258 10.5658	0.1300 8.6781	0.1334 7.4959	0.1363 6.6933
9	0.0735 24091	0.0989 199.4	0.1147 43.8824	0.1257 21.1391	0.1338 13.7716	0.1402 10.3915	0.1452 8.5138	0.1494 7.3386	0.1529 6.5411
10	0.0780 24225	0.1061 199.4	0.1238 43.6858	0.1362 20.9667	0.1455 13.6182	0.1528 10.2500	0.1587 8.3803	0.1635 7.2106	0.1676 6.4172
12	0.0851 24426	0.1175 199.4	0.1384 43.3874	0.1533 20.7047	0.1647 13.3845	0.1737 10.0343	0.1810 8.1764	0.1871 7.0149	0.1922 6.2274
15	0.0926 24630	0.1299 199.4	0.1544 43.0847	0.1723 20.4383	0.1861 13.1463	0.1972 9.8140	0.2063 7.9678	0.2139 6.8143	0.2204 6.0325
20	0.1006 24836	0.1431 199.4	0.1719 42.7775	0.1933 20.1673	0.2100 12.9035	0.2236 9.5888	0.2349 7.7540	0.2445 6.6082	0.2528 5.8318
25	0.1055 24960	0.1516 199.5	0.1831 42.5910	0.2068 20.0024	0.2256 12.7554	0.2410 9.4511	0.2538 7.6230	0.2648 6.4817	0.2744 5.7084
30	0.1089 25044	0.1574 199.5	0.1909 42.4658	0.2163 19.8915	0.2365 12.6556	0.2532 9.3582	0.2673 7.5345	0.2793 6.3961	0.2898 5.6248
60	0.1177 25253	0.1726 199.5	0.2115 42.1494	0.2416 19.6107	0.2660 12.4024	0.2864 9.1219	0.3038 7.3088	0.3190 6.1772	0.3324 5.4104
120	0.1223 25359	0.1805 199.5	0.2224 41.9895	0.2551 19.4684	0.2818 12.2737	0.3044 9.0015	0.3239 7.1933	0.3410 6.0649	0.3561 5.3001
∞	0.1269 25465	0.1887 199.5	0.2337 41.8283	0.2692 19.3247	0.2985 12.1435	0.3235 8.8793	0.3452 7.0760	0.3644 5.9506	0.3815 5.1875

Table C.4 *continued.*

r	c: 10	12	15	20	25	30	60	120	∞
1	4.1E-05 12.8265	4.1E-05 11.7543	4.1E-05 10.7981	4.0E-05 9.9440	4.0E-05 9.4753	4.0E-05 9.1797	4.0E-05 8.4946	3.9E-05 8.1788	3.9E-05 7.8794
2	0.005015 9.4270	0.005015 8.5096	0.005014 7.7008	0.005014 6.9865	0.005014 6.5982	0.005013 6.3547	0.005013 5.7950	0.005013 5.5393	0.005013 5.2983
3	0.0229 8.0807	0.0230 7.2258	0.0232 6.4760	0.0234 5.8177	0.0235 5.4615	0.0235 5.2388	0.0237 4.7290	0.0238 4.4972	0.0239 4.2794
4	0.0477 7.3428	0.0483 6.5211	0.0489 5.8029	0.0496 5.1743	0.0500 4.8351	0.0503 4.6234	0.0510 4.1399	0.0514 3.9207	0.0517 3.7151
5	0.0734 6.8724	0.0747 6.0711	0.0761 5.3721	0.0775 4.7616	0.0784 4.4327	0.0790 4.2276	0.0806 3.7600	0.0815 3.5482	0.0823 3.3499
6	0.0976 6.5446	0.0997 5.7570	0.1019 5.0708	0.1043 4.4721	0.1058 4.1500	0.1069 3.9492	0.1096 3.4918	0.1111 3.2849	0.1126 3.0913
7	0.1193 6.3025	0.1223 5.5245	0.1255 4.8473	0.1290 4.2569	0.1312 3.9394	0.1327 3.7416	0.1368 3.2911	0.1390 3.0874	0.1413 2.8968
8	0.1387 6.1159	0.1426 5.3451	0.1468 4.6744	0.1513 4.0900	0.1543 3.7758	0.1563 3.5801	0.1619 3.1344	0.1649 2.9330	0.1681 2.7444
9	0.1558 5.9676	0.1606 5.2021	0.1658 4.5364	0.1715 3.9564	0.1752 3.6447	0.1778 3.4505	0.1848 3.0083	0.1887 2.8083	0.1928 2.6210
10	0.1710 5.8467	0.1766 5.0855	0.1828 4.4235	0.1896 3.8470	0.1941 3.5370	0.1972 3.3440	0.2058 2.9042	0.2105 2.7052	0.2156 2.5188
12	0.1966 5.6613	0.2038 4.9062	0.2118 4.2497	0.2208 3.6779	0.2267 3.3704	0.2309 3.1787	0.2425 2.7419	0.2491 2.5439	0.2562 2.3583
15	0.2261 5.4707	0.2353 4.7213	0.2457 4.0698	0.2576 3.5020	0.2655 3.1963	0.2712 3.0057	0.2873 2.5705	0.2965 2.3727	0.3067 2.1868
20	0.2599 5.2740	0.2719 4.5299	0.2856 3.8826	0.3014 3.3178	0.3123 3.0133	0.3202 2.8230	0.3429 2.3872	0.3564 2.1881	0.3717 1.9998
25	0.2827 5.1528	0.2967 4.4115	0.3129 3.7662	0.3319 3.2025	0.3451 2.8981	0.3548 2.7076	0.3833 2.2698	0.4006 2.0686	0.4208 1.8771
30	0.2990 5.0706	0.3146 4.3309	0.3327 3.6867	0.3542 3.1234	0.3693 2.8187	0.3805 2.6278	0.4141 2.1874	0.4348 1.9840	0.4596 1.7891
60	0.3443 4.8592	0.3647 4.1229	0.3890 3.4803	0.4189 2.9159	0.4406 2.6088	0.4572 2.4151	0.5096 1.9622	0.5452 1.7469	0.5922 1.5325
120	0.3697 4.7501	0.3931 4.0149	0.4215 3.3722	0.4570 2.8058	0.4834 2.4961	0.5040 2.2998	0.5725 1.8341	0.6229 1.6055	0.6988 1.3637
∞	0.3970 4.6385	0.4240 3.9039	0.4573 3.2602	0.5000 2.6904	0.5327 2.3765	0.5590 2.1760	0.6525 1.6885	0.7333 1.4311	1.0000 1.0000

Bibliography

[1] E. ADAMEC AND R. K. BURDICK, *Confidence intervals for a discrimination ratio in a gauge R&R study with three random factors*, Quality Engineering, 15(3) (2003), pp. 383–389.

[2] C. ARTEAGA, S. JEYARATNAM, AND F. A. GRAYBILL, *Confidence intervals for proportions of total variance in the two-way cross component of variance model*, Communications in Statistics: Theory and Methods, 11 (1982), pp. 1643–1658.

[3] AUTOMOTIVE INDUSTRY ACTION GROUP, *Measurement Systems Analysis*, third edition, AIAG, Detroit, 2002.

[4] C. M. BORROR, D. C. MONTGOMERY, AND G. C. RUNGER, *Confidence intervals for variance components from gauge capability studies*, Quality and Reliability Engineering International, 13 (1997), pp. 361–369.

[5] B. D. BURCH AND H. K. IYER, *Exact confidence intervals for a variance ratio (or heritability) in a mixed linear model*, Biometrics, 53 (1997), pp. 1318–1333.

[6] R. K. BURDICK, A. E. ALLEN, AND G. A. LARSEN, *Comparing variability of two measurement processes using R&R studies*, Journal of Quality Technology, 34 (2002), pp. 97–105.

[7] R. K. BURDICK, N. J. BIRCH, AND F. A. GRAYBILL, *Confidence intervals on measures of variability in an unbalanced two-fold nested design with equal subsampling*, Journal of Statistical Computation and Simulation, 25 (1986), pp. 259–272.

[8] R. K. BURDICK, C. M. BORROR, AND D. C. MONTGOMERY, *A review of methods for measurement systems capability analysis*, Journal of Quality Technology, 35 (2003), pp. 342–354.

[9] R. K. BURDICK AND J. EICKMAN, *Confidence intervals on the among group variance component in the unbalanced one-fold nested design*, Journal of Statistical Computation and Simulation, 26 (1986), pp. 205–219.

[10] R. K. BURDICK AND F. A. GRAYBILL, *Confidence Intervals on Variance Components*, Marcel Dekker, New York, 1992.

[11] R. K. BURDICK AND G. A. LARSEN, *Confidence intervals on measures of variability in R&R studies*, Journal of Quality Technology, 29 (1997), pp. 261–273.

[12] R. K. Burdick, F. Maqsood, and F. A. Graybill, *Confidence intervals on the intraclass correlation in the unbalanced one-way classification*, Communications in Statistics: Theory and Methods, 15 (1986), pp. 3353–3378.

[13] R. K. Burdick, Y.-J. Park, D. C. Montgomery, and C. M. Borror, *Confidence intervals for misclassification rates in a gauge R&R study*, Journal of Quality Technology, to appear in 2005.

[14] A. K. L. Chiang, *A simple general method for constructing confidence intervals for functions of variance components*, Technometrics, 43 (2001), pp. 356–367.

[15] A. K. L. Chiang, *Improved confidence intervals for a ratio in an R&R study*, Communications in Statistics: Simulation, 31 (2002), pp. 329–344.

[16] W. G. Cochran, *Testing a linear relation among variances*, Biometrics, 7 (1951), pp. 17–32.

[17] L. Daniels, R. K. Burdick, and J. Quiroz, *Confidence intervals in a gauge R&R study with fixed operators*, Journal of Quality Technology, to appear in 2005.

[18] K. K. Dolezal, R. K. Burdick, and N. J. Birch, *Analysis of a two-factor R&R study with fixed operators*, Journal of Quality Technology, 30 (1998), pp. 163–170.

[19] A. Donner and G. Wells, *A comparison of confidence interval methods for the intraclass correlation coefficient*, Biometrics, 42 (1986), pp. 401–412.

[20] M. Y. El-Bassiouni, *Short confidence intervals for variance components*, Communications in Statistics: Theory and Methods, 23 (1994), pp. 1915–1933.

[21] M. Y. El-Bassiouni and M. E. M. Abdelhafez, *Interval estimation of the mean in a two-stage nested model*, Journal of Statistical Computation and Simulation, 67 (2000), pp. 333–350.

[22] L. Gong, R. K. Burdick, and J. Quiroz, *Confidence intervals for unbalanced two-factor gauge R&R studies*, Quality and Reliability Engineering International, to appear in 2005.

[23] F. A. Graybill, *Theory and Application of the Linear Model*, Duxbury Press, North Scituate, MA, 1976.

[24] F. A. Graybill and C. M. Wang, *Confidence intervals for proportions of variability in two-factor nested variance component models*, Journal of the American Statistical Association, 74 (1979), pp. 368–374.

[25] F. A. Graybill and C. M. Wang, *Confidence intervals on nonnegative linear combinations of variances*, Journal of the American Statistical Association, 75 (1980), pp. 869–873.

[26] R. Gui, F. A. Graybill, R. K. Burdick, and N. Ting, *Confidence intervals on ratios of linear combinations for non-disjoint sets of expected mean squares*, Journal of Statistical Planning and Inference, 48 (1995), pp. 215–227.

[27] M. HAMADA AND S. WEERAHANDI, *Measurement system assessment via generalized inference*, Journal of Quality Technology, 32 (2000), pp. 241–253.

[28] D. A. HARVILLE AND A. P. FENECH, *Confidence intervals for a variance ratio, or for heritability, in an unbalanced mixed linear model*, Biometrics, 41 (1985), pp. 137–152.

[29] R. P. HERNANDEZ AND R. K. BURDICK, *Confidence intervals and tests of hypothesis on variance components in an unbalanced two-factor crossed design with interaction*, Journal of Statistical Computation and Simulation, 47 (1993), pp. 67–77.

[30] R. P. HERNANDEZ, R. K. BURDICK, AND N. J. BIRCH, *Confidence intervals and tests of hypotheses on variance components in an unbalanced two-fold nested design*, Biometrical Journal, 34 (1992), pp. 387–402.

[31] R. R. HOCKING, *Methods and Applications of Linear Models: Regression and the Analysis of Variance*, second edition, John Wiley & Sons, Hoboken, NJ, 2003.

[32] R. E. HOUF AND D. B. BERMAN, *Statistical analysis of power module thermal test equipment performance*, IEEE Transactions on Components, Hybrids, and Manufacturing Technology, 11 (1988), pp. 516–520.

[33] H. K. IYER AND T. MATHEW, *Comments on Chiang* (2001), Technometrics, 44 (2002), pp. 284–285.

[34] H. K. IYER AND P. PATTERSON, *A Recipe for Constructing Generalized Pivotal Quantities and Generalized Confidence Intervals*. Technical report 2002-10, Department of Statistics, Colorado State University, Fort Collins, 2002. Also available at http://www.stat.colostate.edu/research/2002_10.pdf.

[35] C. R. JENSEN, *Variance component calculations: Common methods and misapplications in the semiconductor industry*, Quality Engineering, 14(4) (2002), pp. 645–657.

[36] M. K. KAZEMPOUR AND F. A. GRAYBILL, *Approximate confidence bounds for ratios of variance components in two-way unbalanced crossed models with interactions*, Communications in Statistics: Simulation and Computation, 20 (1991), pp. 955–967.

[37] L. R. LAMOTTE, *Invariant quadratic estimators in the random, one-way ANOVA model*, Biometrics, 32 (1976), pp. 793–804.

[38] L. R. LAMOTTE, A. MCWHORTER, JR., AND R. A. PRASAD, *Confidence intervals and tests on the variance ratio in random models with two variance components*, Communications in Statistics: Theory and Methods, 17 (1988), pp. 1135–1164.

[39] G. A. LARSEN, *Measurement system analysis in a production environment with multiple test parameters*, Quality Engineering, 16 (2003/2004), pp. 297–306.

[40] J. LEE AND A. I. KHURI, *Comparison of confidence intervals on the among-group variance component for the unbalanced one-way random model*, Communications in Statistics: Simulation and Computation, 31 (2002), pp. 35–47.

[41] Y. LEE AND J. SEELY, *Computing the Wald interval for a variance ratio*, Biometrics, 52 (1996), pp. 1486–1491.

[42] R. A. LEIVA AND F. A. GRAYBILL, *Confidence intervals for variance components in the balanced two-way model with interaction*, Communications in Statistics: Simulation and Computation, 15 (1986), pp. 301–322.

[43] T.-H. LIN AND D. A. HARVILLE, *Some alternatives to Wald's confidence interval and test*, Journal of the American Statistical Association, 86 (1991), pp. 179–187.

[44] S. L. LOHR, *Sampling: Design and Analysis*, Brooks/Cole, Pacific Grove, CA, 1999.

[45] T.-F. C. LU, F. A. GRAYBILL, AND R. K. BURDICK, *Confidence intervals on the ratio of expected mean squares* $(\theta_1 + d\theta_2)/\theta_3$, Biometrics, 43 (1987), pp. 535–543.

[46] T.-F. C. LU, F. A. GRAYBILL, AND R. K. BURDICK, *Confidence intervals on a difference of expected mean squares*, Journal of Statistical Planning and Inference, 18 (1988), pp. 35–43.

[47] T.-F. C. LU, F. A. GRAYBILL, AND R. K. BURDICK, *Confidence intervals on the ratio of expected mean squares* $(\theta_1 - d\theta_2)/\theta_3$, Journal of Statistical Planning and Inference, 21 (1989), pp. 179–190.

[48] D. P. MADER, J. PRINS, AND R. E. LAMPE, *The economic impact of measurement error*, Quality Engineering, 11 (1999), pp. 563–574.

[49] K. D. MAJESKE AND R. W. ANDREWS, *Evaluating measurement systems and manufacturing processes using three quality measures*, Quality Engineering, 15 (2002/2003), pp. 243–251.

[50] G. A. MILLIKEN AND D. E. JOHNSON, *Analysis of Messy Data: Designed Experiments*, Volume 1, Lifetime Learning, Belmont, CA, 1984.

[51] D. C. MONTGOMERY, *Design and Analysis of Experiments*, fifth edition, John Wiley & Sons, New York, 2001.

[52] D. C. MONTGOMERY, *Introduction to Statistical Quality Control*, fourth edition, John Wiley & Sons, New York, 2001.

[53] D. C. MONTGOMERY AND G. C. RUNGER, *Gauge capability and designed experiments. Part I: Basic methods*, Quality Engineering, 6 (1993/1994), pp. 115–135.

[54] A. OLSEN, J. SEELY, AND D. BIRKES, *Invariant quadratic unbiased estimation for two variance components*, Annals of Statistics, 4 (1976), pp. 878–890.

[55] K. PAARK AND R. K. BURDICK, *Confidence intervals for the mean in a balanced two-factor random effect model*, Communications in Statistics: Theory and Methods, 27 (1998), pp. 2807–2825.

[56] D. J. PARK AND R. K. BURDICK, *Performance of confidence intervals in regression models with unbalanced one-fold nested error structures*, Communications in Statistics-Simulation and Computation, 32 (2003), pp. 717–732.

[57] F. E. SATTERTHWAITE, *Synthesis of variance*, Psychometrika, 6 (1941), pp. 309–316.

[58] F. E. SATTERTHWAITE, *An approximate distribution of estimates of variance components*, Biometrics Bulletin, 2 (1946), pp. 110–114.

[59] J. F. SEELY AND Y. EL-BASSIOUNI, *Applying Wald's variance component test*, Annals of Statistics, 11 (1983), pp. 197–201.

[60] J. F. SEELY AND Y. LEE, *A note on the Satterthwaite confidence interval for a variance*, Communications in Statistics: Theory and Methods, 23 (1994), pp. 859–869.

[61] W. J. SPENCER AND P. A. TOBIAS, *Statistics in the semiconductor industry: A competitive necessity*, The American Statistician, 49 (1995), pp. 245–249.

[62] J. D. THOMAS AND R. A. HULTQUIST, *Interval estimation for the unbalanced case of the one-way random effects model*, Annals in Statistics, 6 (1978), pp. 582–587.

[63] N. TING, R. K. BURDICK, AND F. A. GRAYBILL, *Confidence intervals on ratios of positive linear combinations of variance components*, Statistics and Probability Letters, 11 (1991), pp. 523–528.

[64] N. TING, R. K. BURDICK, F. A. GRAYBILL, S. JEYARATMAN, AND T.-F. C. LU, *Confidence intervals on linear combinations of variance components that are unrestricted in sign*, Journal of Statistical Computation and Simulation, 35 (1990), pp. 135–143.

[65] K. TSUI AND S. WEERAHANDI, *Generalized p-values in significance testing of hypotheses in the presence of nuisance parameters*, Journal of the American Statistical Association, 84 (1989), pp. 602–607. (Corrections: 86 (1991), p. 256.)

[66] J. W. TUKEY, *Components in regression*, Biometrics, 7 (1951), pp. 33–69.

[67] S. B. VARDEMAN AND E. S. VANVALKENBURG, *Two-way random-effects analyses and gauge R&R studies*, Technometrics, 41 (1999), pp. 202–211.

[68] A. WALD, *A note on the analysis of variance with unequal class frequencies*, Annals of Mathematical Statistics, 11 (1940), pp. 96–100.

[69] A. WALD, *A note on regression analysis*, Annals of Mathematical Statistics, 18 (1947), pp. 586–589.

[70] C. M. WANG AND F. A. GRAYBILL, *Confidence intervals on a ratio of variances in the two-factor nested components of variance model*, Communications in Statistics-Theory and Methods, A10 (1981), pp. 1357–1368.

[71] S. WEERAHANDI, *Generalized confidence intervals*, Journal of the American Statistical Association, 88 (1993), pp. 899–905. (Corrections: 89 (1994), p. 726.)

[72] S. WEERAHANDI, *Exact Statistical Methods for Data Analysis*, Springer-Verlag, New York, 1995.

[73] J. S. WILLIAMS, *A confidence interval for variance components*, Biometrika, 49 (1962), pp. 278–281.

Index